Methods and Protocols in Food Science

For further volumes:
http://www.springer.com/series/16556

Methods and Protocols in Food Science series is devoted to the publication of research protocols and methodologies in all fields of food science.

Volumes and chapters will be organized by field and presented in such way that the readers will be able to reproduce the experiments in a step-by-step style. Each protocol will be characterized by a brief introductory section, followed by a short aims section, in which the precise purpose of the protocol will be clarified.

Detection and Quantification of Bacteria, Yeast, Viruses, and Protozoan in Foods and Freshwater

Second Edition

Edited by

Marciane Magnani

Department of Food Engineering, Federal University of Paraíba, João Pessoa, Paraíba, Brazil

Geany Targino de Souza Pedrosa

Department of Food Engineering, Federal University of Paraíba, Joao Pessoa, Paraíba, Brazil

Humana Press

Editors
Marciane Magnani
Department of Food Engineering
Federal University of Paraíba
João Pessoa, Paraíba, Brazil

Geany Targino de Souza Pedrosa
Department of Food Engineering
Federal University of Paraíba
Joao Pessoa, Paraíba, Brazil

ISSN 2662-950X ISSN 2662-9518 (electronic)
Methods and Protocols in Food Science
ISBN 978-1-0716-5063-9 ISBN 978-1-0716-5064-6 (eBook)
https://doi.org/10.1007/978-1-0716-5064-6

This Humana imprint is published by the registered company Springer Science+Business Media, LLC, part of Springer Nature.
The registered company address is: 1 New York Plaza, New York, NY 10004, U.S.A.

Preface to the Series

Methods and Protocols in Food Science series is devoted to the publication of research protocols and methodologies in all fields of food science. The series is unique as it includes protocols developed, validated and used by food and related scientists as well as theoretical basis are provided for each protocol. Aspects related to improvements in the protocols, adaptations and further developments in the protocols may also be approached.

Methods and Protocols in Food Science series aims to bring the most recent developments in research protocols in the field as well as very well established methods. As such the series targets undergraduate, graduate and researchers in the field of food science and correlated areas. The protocols documented in the series will be highly useful for scientific inquiries in the field of food sciences, presented in such way that the readers will be able to reproduce the experiments in a step-by-step style.

Each protocol will be characterized by a brief introductory section, followed by a short aims section, in which the precise purpose of the protocol is clarified. Then, an in-depth list of materials and reagents required for employing the protocol is presented, followed by a comprehensive and step-by-step procedures on how to perform that experiment. The next section brings the do's and don'ts when carrying out the protocol, followed by the main pitfalls faced and how to troubleshoot them. Finally, template results will be presented and their meaning/conclusions addressed.

The *Methods and Protocols in Food Science* series will fill an important gap, addressing a common complain of food scientists, regarding the difficulties in repeating experiments detailed in scientific papers. With this, the series has a potential to become a reference material in food science laboratories of research centers and Universities throughout the world.

Campinas, Brazil ***Anderson S. Sant'Ana***

Preface

This book, *Detection and Quantification of Bacteria, Yeast, Viruses, and Protozoan in Foods and Freshwater*, second edition, is a thoroughly updated and enhanced new edition of *Methods and Protocols in Food Science* described in detail by a renowned author team. The detection and quantification of microorganisms in foods and water continue to improve as new matrices are analyzed and methods and protocols are developed. Various protocols have been adapted to reduce analytical time and improve sensitivity and specificity compared to official methods. However, technical errors or minor modifications may alter their accuracy.

This second edition provides a detailed and updated collection of methods and protocols for the detection and quantification of bacteria, viruses, and protozoan in different food matrices (fresh or processed) and freshwater. Particular attention is devoting to virus detection in foods considering the challenges of protocols on this topic. Each chapter introduces the topic, materials, methods, a step-by-step description, figures, graphs, or tables. Tips to avoid mistakes that compromise the analytical technique performance and reduce the working time beyond the different supplies used are also reported. At the end of each chapter, the key references are listed. The book provides basic and comprehensive information about protocols and methods for detection, quantification, and characterization of microorganisms in foods, without disregarding the limitations or details in each specific procedure.

João Pessoa, Paraíba, Brazil — *Marciane Magnani*
João Pessoa, Paraíba, Brazil — *Geany Targino de Souza Pedrosa*

Contents

Contributors

JORGE ANDALUZ-ARBE • *Departamento de Producción Animal y Ciencia de los Alimentos, Facultad de Veterinaria, Instituto Agroalimentario de Aragón-IA2 (Universidad de Zaragoza-CITA), Zaragoza, Spain*

LIVIA BALAGUER • *Departamento de Producción Animal y Ciencia de los Alimentos, Facultad de Veterinaria, Instituto Agroalimentario de Aragón-IA2 (Universidad de Zaragoza-CITA), Zaragoza, Spain*

SUSANA BAYARRI • *Departamento de Producción Animal y Ciencia de los Alimentos, Facultad de Veterinaria, Instituto Agroalimentario de Aragón-IA2 (Universidad de Zaragoza-CITA), Zaragoza, Spain*

DANIEL BERDEJO • *Departamento de Producción Animal y Ciencia de los Alimentos, Facultad de Veterinaria, Instituto Agroalimentario de Aragón-IA2 (Universidad de Zaragoza-CITA), Zaragoza, Spain*

RAÚL CAMPILLO • *Departamento de Producción Animal y Ciencia de los Alimentos, Facultad de Veterinaria, Instituto Agroalimentario de Aragón-IA2 (Universidad de Zaragoza-CITA), Zaragoza, Spain*

MARIE-ÈVE COLLARD • *Department of Food Science, Institute of Nutrition and Functional Foods, Université Laval, Québec City, QC, Canada*

MARÍA PILAR CONCHELLO • *Departamento de Producción Animal y Ciencia de los Alimentos, Facultad de Veterinaria, Instituto Agroalimentario de Aragón -IA2-, Universidad de Zaragoza-CITA, Zaragoza, Spain*

BERDEJO DANIEL • *Departamento de Producción Animal y Ciencia de los Alimentos, Facultad de Veterinaria, Instituto Agroalimentario de Aragón-IA2 (Universidad de Zaragoza-CITA), Zaragoza, Spain*

MERIANE DEMOLINER • *Molecular Microbiology Laboratory, Feevale University, Novo Hamburgo, RS, Brazil; Laboratório de Microbiologia Molecular, Universidade Feevale, Novo Hamburgo, RS, Brazil*

GARCÍA-GONZALO DIEGO • *Departamento de Producción Animal y Ciencia de los Alimentos, Facultad de Veterinaria, Instituto Agroalimentario de Aragón-IA2 (Universidad de Zaragoza-CITA), Zaragoza, Spain*

LAURA ESPINA • *Departamento de Producción Animal y Ciencia de los Alimentos, Facultad de Veterinaria, Instituto Agroalimentario de Aragón-IA2 (Universidad de Zaragoza-CITA), Zaragoza, Spain; Fundación Agencia Aragonesa para la Investigación y el Desarrollo (ARAID), Zaragoza, Spain*

ALBERTO FAU • *Departamento de Producción Animal y Ciencia de los Alimentos, Facultad de Veterinaria, Instituto Agroalimentario de Aragón-IA2 (Universidad de Zaragoza-CITA), Zaragoza, Spain*

DIEGO GARCÍA-GONZALO • *Departamento de Producción Animal y Ciencia de los Alimentos, Facultad de Veterinaria, Instituto Agroalimentario de Aragón-IA2 (Universidad de Zaragoza-CITA), Zaragoza, Spain*

Ivo García-Penas • *Departamento de Producción Animal y Ciencia de los Alimentos, Facultad de Veterinaria, Instituto Agroalimentario de Aragón-IA2 (Universidad de Zaragoza-CITA), Zaragoza, Spain*
Nerea Garín-Murguialday • *Centro Nacional de Tecnología y Seguridad Alimentaria (CNTA), San Adrián, Spain*
Diego Gómez-Lozano • *Departamento de Producción Animal y Ciencia de los Alimentos, Facultad de Veterinaria, Instituto Agroalimentario de Aragón-IA2 (Universidad de Zaragoza-CITA), Zaragoza, Spain*
Juliana Schons Gularte • *Molecular Microbiology Laboratory, Feevale University, Novo Hamburgo, RS, Brazil; Immunology and Molecular Biology Laboratory, School of Veterinary Medicine, Federal University of Rio Grande do Sul, Porto Alegre, RS, Brazil; Laboratório de Microbiologia Molecular, Universidade Feevale, Novo Hamburgo, RS, Brazil; Laboratório de Imunologia e Biologia Molecular, Faculdade de Veterinária, Universidade Federal Do Rio Grande Do Sul, Porto Alegre, RS, Brazil*
Nabil Halaihel • *Departamento de Patología Animal, Facultad de Veterinaria, Universidad de Zaragoza, Zaragoza, Spain*
Julie Jean • *Department of Food Science, Institute of Nutrition and Functional Foods, Université Laval, Québec City, QC, Canada; Faculty of Agriculture and Food Sciences, Department of Food Sciences, Institute of Nutrition and Functional Foods, Université Laval Pavillon Paul-Comtois, Québec City, QC, Canada*
Éric Jubinville • *Department of Food Science, Institute of Nutrition and Functional Foods, Université Laval, Québec City, QC, Canada*
Espina Laura • *Fundación Agencia Aragonesa para la Investigación y el Desarrollo (ARAID), Zaragoza, Spain*
Anne-Marie Lauzier • *Department of Food Science, Institute of Nutrition and Functional Foods, Université Laval, Québec City, QC, Canada*
Regina Lázaro • *Departamento de Producción Animal y Ciencia de los Alimentos, Facultad de Veterinaria, Instituto Agroalimentario de Aragón-IA2 (Universidad de Zaragoza-CITA), Zaragoza, Spain*
Graciela Volz Lopes • *Department of Agroindustrial Science and Technology, Eliseu Maciel Faculty of Agronomy, Federal University of Pelotas, University Campus, Capão do Leão-RS, Brazil*
Marciane Magnani • *Laboratory of Microbial Processes in Foods, Department of Food Engineering, Technology Center, Federal University of Paraíba, João Pessoa, Brazil*
Natalia Merino • *Departamento de Producción Animal y Ciencia de los Alimentos, Facultad de Veterinaria, Instituto Agroalimentario de Aragón-IA2 (Universidad de Zaragoza-CITA), Zaragoza, Spain*
Rafael Pagán • *Departamento de Producción Animal y Ciencia de los Alimentos, Facultad de Veterinaria, Instituto Agroalimentario de Aragón-IA2 (Universidad de Zaragoza-CITA), Zaragoza, Spain*
Edson Douglas Silva Pontes • *Laboratory of Microbial Processes in Foods, Department of Food Engineering, Technology Center, Federal University of Paraíba, João Pessoa, Brazil*
Pagán Rafael • *Departamento de Producción Animal y Ciencia de los Alimentos, Facultad de Veterinaria, Instituto Agroalimentario de Aragón-IA2 (Universidad de Zaragoza-CITA), Zaragoza, Spain*
Belén Romero de Castilla • *Departamento de Producción Animal y Ciencia de los Alimentos, Facultad de Veterinaria, Instituto Agroalimentario de Aragón -IA2-, Universidad de Zaragoza-CITA, Zaragoza, Spain*

MARÍA CARMEN ROTA • *Departamento de Producción Animal y Ciencia de los Alimentos, Facultad de Veterinaria, Instituto Agroalimentario de Aragón -IA2-, Universidad de Zaragoza-CITA, Zaragoza, Spain*
RUTHCHELLY TAVARES DA SILVA • *Laboratory of Microbial Processes in Foods, Department of Food Engineering, Technology Center, Federal University of Paraíba, João Pessoa, Brazil*
DÉBORA RODRIGUES SILVEIRA • *Department of Agroindustrial Science and Technology, Eliseu Maciel Faculty of Agronomy, Federal University of Pelotas, University Campus, Capão do Leão-RS, Brazil*
FERNANDO ROSADO SPILKI • *Molecular Microbiology Laboratory, Feevale University, Novo Hamburgo, RS, Brazil; Laboratório de Microbiologia Molecular, Universidade Feevale, Novo Hamburgo, RS, Brazil*
RAQUEL VIRTO • *Centro Nacional de Tecnología y Seguridad Alimentaria (CNTA), San Adrián, Spain*

Chapter 1

Sampling and Analysis of Cheese for Antimicrobial Resistance Characterization and Isolation of Target Bacteria

Livia Balaguer, Natalia Merino, García-Gonzalo Diego, Pagán Rafael, Berdejo Daniel and Espina Laura

Abstract

Cheese, a popular dairy product, is produced through the microbial fermentation of milk, resulting in a product with a diverse microbiological profile shaped by various influencing factors. Bacteria in cheese contribute to its physicochemical and sensory profile, but some also might play a critical role in antimicrobial resistance (AMR). Due to antibiotic use in dairy farming, bacterial strains in cheese can harbour antibiotic resistance genes (ARGs) transferable to human pathogens, generating AMR risks. Raw milk cheeses may be particularly notable as reservoirs with potential AMR risks. This chapter outlines a culture-based method to assess the microbiological composition and antibiotic resistance profiles of cheese samples and the use of Sanger sequencing for characterising some bacteria of interest.

Keywords Cheese, Microbiome, Resistome, Antibiotic resistant bacteria, AMR, Culture-dependent techniques, PCR, Sanger sequencing

1 Introduction

Cheese is a commonly consumed dairy product with a history of production dating back thousands of years. Cheese is produced from milk through a combination of microbial fermentation and enzymatic processes, and its microbiological content can be derived from either the initial microbiota of the milk, added starter cultures, or a combination of both. The final microbial profile depends on the type of milk used, the production environment, the fermentation process, and the ripening process [1].

While microbial species in cheese contributing to its sensory characteristics encompass both fungi and bacteria, the latter group is of especial relevance due to their primary role in antimicrobial resistance (AMR) [2, 3]. The inappropriate use of antibiotics in animal husbandry, including dairy farming, has accelerated the emergence of resistant bacterial strains in food products [4]. These bacteria may carry

Marciane Magnani and Geany Targino de Souza Pedrosa (eds.), *Detection and Quantification of Bacteria, Yeast, Viruses, and Protozoan in Foods and Freshwater*, Methods and Protocols in Food Science, https://doi.org/10.1007/978-1-0716-5064-6_1,

antibiotic resistance genes (ARGs) that can be transferred to human pathogens, further exacerbating the AMR crisis [5]. Lactic acid bacteria (LAB), such as *Lactobacillus* spp. and *Streptococcus* spp., play a key role in cheese fermentation, and other bacteria, including *Staphylococcus* spp., *Enterococcus* spp., and coliforms, can also be present. All of these species are capable of carrying ARGs [6]. Therefore, understanding the microbial communities in cheese and assessing their antibiotic resistance profiles is critical for ensuring food safety and preventing the spread of resistant bacteria.

Raw milk cheeses, in particular, have been identified as potential reservoirs of antibiotic-resistant bacteria due to the natural microbiota and their minimal processing. Studies have shown that bacterial isolates from these cheeses can harbor resistance to common antibiotics, including penicillin, tetracycline, and macrolides. The potential for these bacteria to act as vectors of antibiotic resistance highlights the importance of screening and characterizing bacterial populations in cheese, not only to ensure food safety, but also to address the broader public health challenge posed by AMR [7–10].

In this chapter, we describe the methodology for the culture-based assessment of the microbiological composition of a cheese sample. This approach provides a fingerprint of each cheese sample in terms of its general microbiota and its possible antibiotic resistance profile and facilitates the isolation and detailed genotypic characterization of potential antibiotic-resistant strains employing Sanger sequencing.

2 Materials

2.1 Sample Collection and Homogenization

1. Clean environment.
2. Sterile knife.
3. Sterile tongs.
4. Analytical balance.
5. Buffered peptone water (BPW).
6. Sterile filter bags to retain particles during sample homogenization.
7. Peristaltic homogenizer (stomacher).
8. Sterile bottles.

2.2 Culturing Method

1. Vortex mixer.
2. Peptone water (PW).
3. Sterile Screw-Cap Tubes.
4. Culturing tools: micropipettes, petri dishes (90 mm), and L-shaped spreaders or glass beads.

5. Growth media prepared according to the manufacturer instructions:
   (a) De Man Rogose Sharpe Agar (MRS) for counting lactic acid bacteria.
   (b) Bile Esculin Agar (BEA) for counting *Enterococcus* spp. and *Streptococcus* group D spp.
   (c) Potato Dextrose Agar (PDA) for counting yeasts and molds.
   (d) Violet Red Bile Glucose Agar (VRBG) for counting.
   (e) Mannitol Salt Agar (MSA) for counting spp., particularly pathogenic strains of *Staphylococcus*.
6. Antibiotics (Tables 1 and 2):
   (a) Ampicillin sodium salt (AMP).
   (b) Chloramphenicol (CHL).
   (c) Erythromycin (ERY).
   (d) Tetracycline (TET).
   (e) Streptomycin sulfate salt (STR).
7. Incubator chamber (25–37°C).
8. Anaerobic jar with gas pack (Thermo Scientific™ Oxoid AnaeroGen™ 2.5 L Bags) or anaerobic incubator.

2.3 Colony-PCR, Sanger Sequencing of the 16S rRNA Gene, and Storage of Isolates

1. Microtubes 1.5 mL.
2. Sterile toothpicks or inoculation needles.
3. Milli-Q sterile water.
4. Microcentrifuge.

Table 1
Diluent and stock solution for each antibiotic (CLSI M100-S24)

			MRS medium	BEA medium
	Diluent	Stock solution (mg/mL)	Final concentration on plate (μg/mL)	
Ampicillin	Phosphate buffer pH = 8; 0.1 mol/L	50	2	2
Chloramphenicol	Ethanol at 95%	50	4	16
Erythromycin	Ethanol at 95%	50	1	4
Tetracycline	H_20	50	8	4
Streptomycin	H_20	50	64	128

Breakpoints defined in terms of the minimum inhibitory concentration (MIC) by EFSA (2012)

Table 2
Example for FoR of *Enterococcus*

Number of bacteria grown in antibiotic agar (log_{10} CFU/g)	Number of bacteria in BEA medium supplemented with antibiotics (log_{10} CFU/g)				
Control	Ampicilin 2 μg/mL	Chloranphenicol 4 μg/mL	Erytromycin 2 μg/mL	Tetracycline 4 μg/mL	Streptomycin 64 μg/mL
5.43	5	2.95	4.88	4.89	2.20
6.43	5.11	3.31	5.86	5.86	2.38
6.02	5.34	3.74	5.37	5.27	3.00
Mean ± standard deviation					
5.98 ± 0.41	5.15 ± 0.17	3.33 ± 0.39	5.37 ± 0.49	5.34 ± 0.49	2.53 ± 0.42
Mean FoR					
	1.61×10^{-1}	**2.24×10^{-3}**	**2.45×10^{-1}**	**2.3×10^{-1}**	**3.55×10^{-4}**

Table 3
Example list of components for each PCR reaction

Reagent	Volume (μL) for 25 μL reaction	Final concentration
10× Reaction Buffer (with $MgCl_2$)	2.5	1×
dNTPs (10 mM)	0.5	200 μM
Forward Primer (10 μM)	1	0.4 μM
Reverse Primer (10 μM)	1	0.4 μM
Template DNA (previous preparation)	5	10–100 ng
Hot-Start Taq DNA polymerase (5 U/ μL)	0.2	1 unit
Nuclease-free water	Added up to a total volume of 25	

5. Thermal cycler.
6. PCR reagents (DNA Polymerase, deoxynucleotides (dNTPs), and buffer solution) (see Table 3).
7. Forward and reverse primers (see Table 4).
8. 0.2-mL PCR tubes.
9. Agar plates, inoculation loops, and pre-filled cryovials for storage of strains.
10. Ultra-low temperature freezer (−80°C).

Table 4
Commonly used primer pairs for 16S rRNA amplification and sequencing

Primer pair	Sequences	Amplified region	Amplicon length (bp)	Reference
27F 1492R	5'-AGAGTTTGATCMTGGCTCAG-3' 5'-TACGGYTACCTTGTTACGACTT-3'	V1–V9	~1500	[12]
27F 534R	5'-AGAGTTTGATCMTGGCTCAG-3' 5'-ATTACCGCGGCTGCTGG-3'	V1–V3	~300	[13]
515F 944R	5'-GTGCCAGCMGCCGCGGTAA-3' 5'-GAATTAAACCACATGCTC-3'	V4–V5	~500	[14]

2.4 Taxonomic Identification of Sequenced Isolates

1. Sequence data.
2. Workstation with access to internet and software to view sequence data (e.g., Unipro UGENE [11]).

3 Methods

3.1 Experimental Design

This study aims to assess the general microbiota and antibiotic resistance profile of ripened cheeses. Figure 1 shows the experimental steps.

3.2 Sample Collection

A representative sample from cheese is necessary for laboratory analysis. 25 g of cheese are needed, taking small pieces from all parts of the cheese while avoiding the rind. For cutting, a knife or guillotine can be used, previously washed with water and soap, bleach, and finally ethanol. This ensures complete sterility. Using sterilized tongs, place the cheese portions in sterile filter bags along with 225 mL of BPW (Fig. 2).

3.3 Sample Homogenization

25 g of cheese are mixed with 225 mL of BPW in filter bags and homogenized at 300 revolutions per minute (rpm) during 5 min in a peristaltic homogenizer. In this way, a stock homogenate of each sample is obtained. The resulting stock dilution is transferred to sterile bottles avoiding the cheese fat (*see* **Note 1**). This homogenate can be used immediately or stored at −20°C for later analysis.

3.4 Culturing Method for General Microbial Count

To assess the present microbiota in the cheeses, the following microbial groups are cultured in different selective and/or differential growth media using an appropriate dilution (*see* **Note 2**): LAB, *Enterococcus* spp., *Enterobacteriaceae*, yeasts, molds, and *Staphylococcus* spp. (Fig. 3).

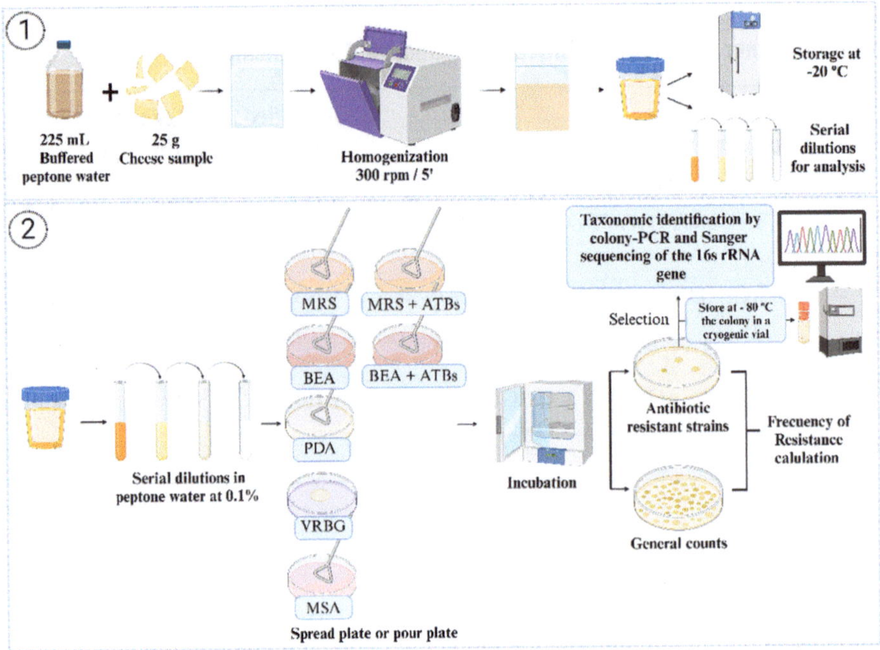

Fig. 1 1. Workflow for sample preparation. 2. Culture-based microbiome and resistome characterization workflow

Fig. 2 Sample collection. 1. Total piece. 2. Cut the cheese in half. 3. Remove the rind. 4. Cut the piece of cheese in small pieces choosing from all parts of cheese. 5. Weigh 25 g of cheese and put it in a filter bag with 225 mL of BPW

LAB are cultured in MRS. For this purpose, 100 μL of stock dilution and, if necessary, of the corresponding decimal dilution are spread-plated (*see* **Note 3**). Once the inoculum has been absorbed (10–15 min), the plates are incubated in inverted position and anaerobic conditions at 30°C during 72 h. After the incubation period, the colonies on each plate are counted.

Enterococcus spp. culturing is carried out using BEA medium. This medium differentiates *Enterococcus* spp. based on their ability to hydrolyze esculin in the presence of bile, a characteristic not common to many other bacteria. 100 μL of stock dilution and, if necessary, of the corresponding decimal dilution are spread-plated (*see* **Note 4**).

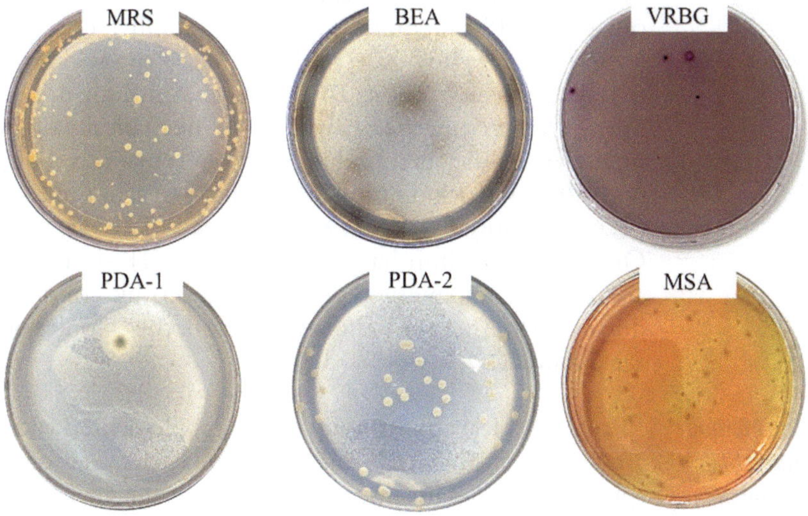

Fig. 3 Morphological diversity among microbial groups grown in different selective growth media (De Man Rogose Sharpe Agar, Bile Esculin Agar, Violet Red Bile Glucose Agar, Potato Dextrose Agar-1 (molds example), Potato Dextrose Agar-2 (yeast example) and Mannitol Salt Agar)

Once the inoculum has been absorbed (10–15 min), the plates are incubated in inverted position and aerobic conditions at 37°C during 48 h. After the incubation period, the colonies on each plate are counted (*see* **Note 5**).

Enterobacteriaceae culturing is carried out following the standard UNE-EN ISO 21528-2:2018. This medium is selective for Gram-negative bacteria, inhibiting Gram-positive organisms. 1 mL of stock dilution and, if necessary, of the corresponding decimal dilution is pour-plated on VRBG (*see* **Note 6**). Once solidified, another layer of VRGB (5–10 mL) is added to prevent wide-spread growth and to archive semi-anaerobic conditions. After solidification of the agar, plates are incubated in inverted position at 37°C during 24 h, afterward, the colonies exhibiting a pink halo on each plate are counted.

Following the standard UNE 34821:1986 for the culturing of yeasts and molds , 1 mL or 100 μL of stock dilution and, if necessary, of the corresponding decimal dilution (*see* **Note** 7), are spread-plated on PDA. To prevent the growth of other microorganisms, PDA medium is prepared following the ISO standard, being supplemented with 100 mg/L of chloramphenicol and 50 mg/L of gentamicin. Once the inoculum has been absorbed (10–15 min), the plates are incubated in inverted position and aerobic conditions at 25°C during 72 h. After the incubation period, the colonies on each plate are counted.

Detection and enumeration of *Staphylococcus* spp. are carried out using MSA medium. Pathogenic Staphylococci colonies will appear with a yellow halo on MSA due to the acid produced from mannitol fermentation. In contrast, although non-pathogenic *Staphylococcus* spp, species will grow, theytypically do not ferment mannitol, so the surrounding medium remains red or pink. 1 mL or 100 μL of stock dilution and, if necessary, of the corresponding decimal dilution (*see* **Note 8**), are spread-plated on MSA medium. Once the inoculum has been absorbed (10–15 min), the plates are incubated in inverted position and aerobic conditions at 37°C during 48 h. After the incubation period, the colonies on each plate are counted.

3.5 Culturing Method for General Microbial Count with Antibiotics

Microbiological counts using media supplemented with antibiotics are a common practice in microbiology to selectively count or isolate specific microorganisms, while inhibiting the growth of unwanted species. This technique is particularly useful when the focus is on specific groups, such as antibiotic-resistant bacteria.

MRS and BEA media are supplemented with antibiotics to isolate antibiotic-resistant bacteria. The European Food Safety Authority (EFSA) has established cut-off values to classify bacteria as susceptible, intermediate, or resistant to specific antibiotics. These breakpoints are based on the Minimum Inhibitory Concentration (MIC), which is the lowest concentration of an antibiotic that inhibits bacterial growth. EFSA uses these values in its guidelines for assessing antimicrobial resistance, particularly in food safety and zoonotic bacteria (Table 1).

To achieve this, 100 μL of stock dilution and, if necessary, of the corresponding decimal dilution (*see* **Note 9**) are spread-plated onto the appropriate growth medium supplemented with each antibiotic (*see* **Note 10**). Once the inoculum has been absorbed (10–15 min), the plates are incubated in inverted position following the same procedure described previously for each medium. After the incubation period, colonies on each plate are counted.

3.6 Assessment of Frequency of Resistance (FoR) in Bacteria

The first step to calculate the Frequency of Resistance (FoR) in bacteria involves determining the original concentration of bacteria in the general count (without exposure to antibiotics). To calculate the concentration of bacteria, the following formula is used:

$$\textit{Total number of bacteria}\left(\frac{CFU}{g}=\frac{CFU}{mL}\right)=\frac{\textit{Number of colonies x dilution factor}}{\textit{Volume plated }(mL)}$$

where:

- **Number of colonies**: the colony count from the plate.

- **Dilution factor**: the reciprocal of the dilution from which the sample was plated (e.g., if plated from a 1:1000 dilution, the dilution factor is 1000). It is necessary to include the first dilution which is obtained when 25 g of cheese is homogenized in 225 mL of BPW.
- **Volume plated**: The amount of sample plated, usually 0.1 mL or 1 mL.

To calculate FoR the following formula is applied (see an example in Table 2):

$$FoR = \frac{Number\ of\ bacteria\ grown\ in\ antibiotic\ agar\ log\ log\ 10\ \left(\frac{CFU}{g}\right)}{Number\ of\ bacteria\ in\ non-antibiotic\ agar\ log\ log\ 10\ \left(\frac{CFU}{g}\right)}$$

3.7 Taxonomic Identification by Colony-PCR, Sanger Sequencing of the 16S rRNA Gene Along with Storage of Isolates for Further Characterization

Once bacterial colonies have grown onto agars of interest, some are collected for taxonomic identification by PCR (Polymerase Chain Reaction), storage in cryovials and further characterization. PCR is a widely used molecular biology technique for amplifying specific DNA sequences. A colony-PCR uses the bacterial colony mass as DNA template, simplifying the process. In bacterial taxonomic identification, a common amplification target is the **16S ribosomal RNA (rRNA)** gene. This gene is a component of the 30S subunit of prokaryotic ribosomes and contains nine hypervariable regions (V1–V9) that can provide species-specific signature sequences, making it an excellent marker for distinguishing bacterial and archaeal species. The resulting amplicon can be sequenced using the Sanger method to determine its nucleotide sequence, enabling precise identification. These steps are outlined below.

3.7.1 Sample Collection and Colony Isolation

The first step is to collect 5 or 10 colonies of interest from the agar plates previously incubated (*see* **Note 11**). Using a sterile toothpick or inoculating needle, a single colony is picked up and transferred into a PCR tube that contains 20 μL of milli-Q sterile water. Additionally, the colony should be isolated onto new agar for future storage in a cryovial (*see* **Note 12**). To improve cell lysis, the microtube is incubated for 10 min at 95°C, and spun down at 11,000 RFC (Relative Centrifugal Force) for 1 min. Then, 5 μL of the resulting supernatant is used as the DNA template for the PCR reaction.

3.7.2 PCR (Polymerase Chain Reaction)

The key components of the PCR are (see Table 3):

1. Template DNA: lysed cell colony as explained above.
2. Primers: there are different primer pairs commonly used to target different regions of the 16S rRNA gene (see Table 4). It is recommended to use the universal primer pair 27F and 1492R as it spans nearly full-length of 16S rRNA gene.

Table 5
Typical cycling conditions for rRNA amplification using Hot-Start Taq DNA polymerase

Cycling step	Temperature (°C)	Holding time	Number of cycles
Enzyme activation	95	2 min	1
Denaturation	95	30 s	35
Annealing	53	30 s	
Extension	72	1.5 min (1 min/kb)	
Final extension	72	10 min	1
Hold	4	Until use	

3. DNA polymerase: different types of DNA polymerases can be used, being Hot-Start Taq DNA polymerase preferred for colony PCR.
4. dNTPs.
5. Buffer solution.
6. Nuclease-free water.

For each PCR reaction, the specific components, their volumes, and the order of addition depend on the specific DNA polymerase used, and following the DNA polymerase's manufacturer instructions is crucial (*see* **Note 13**).

The thermal cycler conditions of the PCR depend on the specific DNA polymerase and primers used in the reaction. In particular, the annealing temperature is chosen based on the melting temperature of the primers, which is determined by their nucleotide sequences. It is always recommended to follow the DNA polymerase manufacturer's instructions (see Table 5 for typical cycling conditions using HotStart Taq DNA polymerase and primers 27F and 1492R).

3.7.3 Purification of PCR Products

Purification of PCR products after sequencing is crucial to remove unwanted components that might interfere with subsequent sequencing. The goal is to eliminate excess primers, dNTPs, salts, and other contaminants. Common methods for PCR product purification include commercial spin column kits, enzymatic treatment (ExoSAP-IT) and magnetic beads, any of which is suitable for this step.

3.7.4 Sanger Sequencing and Taxonomic Identification

Sanger sequencing is a precise method used to determine the nucleotide sequence of a DNA fragment. It involves using DNA polymerase to replicate the target DNA in a solution containing both standard nucleotides and special chain-terminating dideoxynucleotides (ddNTPs), each tagged with a fluorescent dye. When a ddNTP is incorporated, it stops DNA elongation, creating fragments of varying

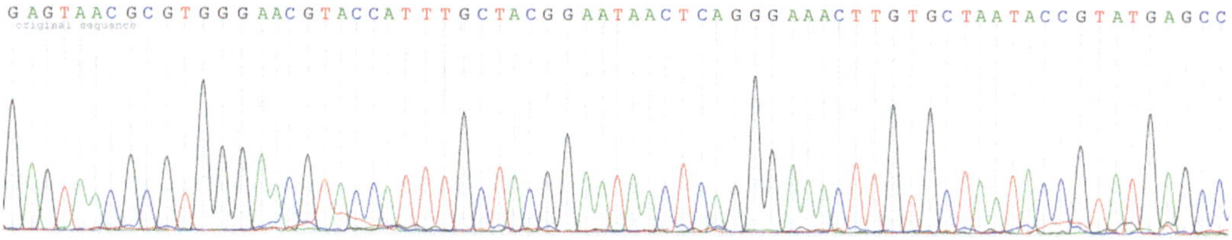

Fig. 4 Example of sequencing chromatogram. This figure shows a representative chromatogram from DNA sequencing, highlighting the peaks corresponding to the base calls in the sequence. The x-axis represents the position along the sequence, and the y-axis shows the intensity of the detected signals for each nucleotide

lengths, each ending at a specific nucleotide. These fragments are then separated by size through capillary electrophoresis, where a laser detects the fluorescent labels, allowing the nucleotide sequence to be identified.

Before starting, it is crucial to know the DNA fragment concentration, typically in the range of 10–20 ng/μL, adjusted for fragment length. Unlike PCR, which uses two primers to amplify both strands, Sanger sequencing requires only one primer, as it sequences a single strand. Therefore, ensure that the sample sent for sequencing includes only one primer. The volume of primer depends on the sanger methodology performed. If possible, it is recommended to send two samples per 16S rRNA, one with the 27F primer and another with the 1492R primer, to help cover the entire 1500 bp and increase the accuracy of the reconstructed sequence.

It is essential to inspect the sequencing chromatogram before validating the DNA sequence. Each base is indicated by a specific color, read in order from left to right and top to bottom, corresponding to the 5′ to 3′ direction of the DNA stand. A high-quality chromatogram will show evenly spaced, single color peak with minimal baseline noise, indicating reliable sequence (see Fig. 4).

The final step is to use the Basic Local Alignment Search Tool (BLAST [15]) to find matches between the obtained nucleotide sequence and those in its rRNA/ITS database. To ensure a meaningful match between sequences, the *e*-value should be very low ($<1^{-5}$). This low *e*-value threshold indicates that the observed alignment is highly significant and unlikely to have occurred by chance. Additionally, query coverage should be ≥90%, meaning that at least 90% of the query sequence aligns with the subject sequence. Regarding the identity, a >99% identity in 16S rRNA gene sequence is the criterion used to identify an isolate to the species level, and a 97–99% identity is used to identify an organism at the genus level [16].

3.7.5 Further Characterization

Once taxonomically identified, bacterial colonies of interest can be regrown from their stored cryovials and subjected for further identification. For example, colonies recovered from agar containing chloramphenicol can be subjected to PCR with primers targeting genes conferring resistance to this antibiotic. The resulting PCR product can then be run on an electrophoresis gel to visualize the presence of the expected fragment, indicating the presence of the targeted resistance gene. Furthermore, to quantify the copy numbers of each gene of interest with high accuracy, qPCR and metagenomic analyses, can be employed. These techniques provide detailed insights into the abundance and functional potential of resistance genes within the bacterial population.

4 Notes

1. A filter bag equipped with a membrane is used to separate the liquid matrix from fat. It is crucial at this stage to ensure that only the liquid phase is collected, avoiding the fat matrix.
2. It is recommended to spread alternate dilutions to achieve an optimal bacterial count when working with microbial cultures. For example, performing dilutions at 10^{-4}–10^{-6} or 10^{-3}–10^{-5} provides a balanced approach. This method ensures that at least one of the plates will have a countable number of colonies within the ideal range.
3. Since LAB are the most representative microbial group in cheese, achieving an optimal count requires diluting the initial sample until reach 10^{-5} dilution. To accomplish this, the sample is diluted in 0.1% PW. The recommended method involves using sterile screw-cap tubes, each filled with 9 mL of 0.1% PW.

   Serial dilution process:

   - Start by adding 1 mL of the stock solution to the first tube.
   - Then perform serial dilutions by transferring 1 mL from the first tube into the second, continuing this process across all tubes to reach the desired dilution factor.

   This approach ensures the correct bacterial count range, allowing for accurate quantification of LAB in the sample when plated on selective media. LAB counts in cheese are typically high, often found in the range of 7–8 logarithmic units (10^7–10^8 colony-forming units per g, CFU/g).
4. *Enterococcus* spp. is the second most representative bacterial species in cheese, so to obtain an optimal count, it is necessary to dilute the sample, like the method outlined for LAB in **Note 3**. It is recommended to dilute the sample until reach 10^{-4} dilution. This dilution step is crucial because *Enterococcus* spp. is generally found in the range of 10^4–10^6 CFU/g.

5. BEA medium its specific for *Enterococcus* spp., but sometimes other similar species can grow. For that reason, it is important to note that *Enterococcus* spp. show a gray halo around them, which is not observed in other bacteria.
6. Enterobacteriaceae are not commonly found in cheese, however, it is a crucial metric in food safety because of the associated risks, so in this case, 1 mL of stock solution is directly added to the growth medium and pour-plated. Since these bacteria are expected to be present in low concentrations, the anticipated count is in the range of 10–100 CFU/g.
7. Yeasts and molds may be present in cheeses and are often found in the range of 10^2–10^3 CFU/g, so it is recommended to dilute the sample. In some cheeses where molds are inoculated as starters, the concentration may range from 10^4 to 10^5 CFU/g.
8. The concentration of *Staphylococcus* spp. in cheese, particularly pathogenic staphylococci, is an important metric in food safety due to the risk of toxin production. It is not common to find this bacterial group, but if it appears, concentrations typically range from 10^3 to 10^5 CFU/g.
9. In that case, the focus is on LAB and *Enterococcus* spp. In general counts around 10^7–10^8 for the first group and 10^4–10^6 CFU/g for the latter would be expected. Counts will be lower in growth media with antibiotics. Therefore, it is recommended to perform 2 fewer dilutions than for general counts.
10. It is important to specify at which step the antibiotic is added. The antibiotic must be sterile and should be added once the medium has been sterilized, and the temperature is around 50°C. This is crucial because if the medium is too hot, the antibiotic may degrade, leading to a loss of activity.
11. At this point, 4 or 5 colonies of interest grown on the surface of agar plates should be selected. It is important that these are well-separated from other colonies.
12. The agar used for colony growing must be the same as previously used. Cryovials allow for the storage of microorganisms at very low temperature, which minimizes metabolic activity and halts cellular growth. This ensures that the colonies are preserved in their original state for extended periods.
13. A "master mix" for PCR simplifies setup by combining essential components (Taq polymerase, dNTPs and buffer) in a single solution, requiring only the addition of DNA and primers. This reduces pipetting steps, minimizes contamination risk, and improves consistency and reproducibility across samples.

References

1. Bansal V, Veena N (2024) Understanding the role of pH in cheese manufacturing: general aspects of cheese quality and safety. J Food Sci Technol 61:16–26
2. WHO (2014) Antimicrobial resistance. Global report on surveillance. World Heal Organ 61:12–28
3. Murray CJ, Ikuta KS, Sharara F, Swetschinski L, Robles Aguilar G, Gray A et al (2022) Global burden of bacterial antimicrobial resistance in 2019: a systematic analysis. Lancet 399:629–655
4. Santamarina-García G, Amores G, Llamazares D, Hernández I, Javier R, Barron L, Virto M (2024) Phenotypic and genotypic characterization of antimicrobial resistances reveals the effect of the production chain in reducing resistant lactic acid bacteria in an artisanal raw ewe milk PDO cheese. Food Res Int 187:114308
5. CDC. Antimicrobial Resistance. https://www.cdc.gov/antimicrobial-resistance/?CDC_AAref_Val=https://www.cdc.gov/drugresistance/pdf/infographic-advances-in-antibiotic-resistance-508c.pdf. Retrieved on 12 November 2025
6. Nalepa B, Markiewicz LH (2022) Microbiological biodiversity of regional cow, goat and ewe milk cheeses produced in poland and antibiotic resistance of lactic acid bacteria isolated from them. Animals. 13:168
7. György É, Unguran KA, Laslo É (2023) Biocide tolerance and impact of sanitizer concentrations on the antibiotic resistance of bacteria originating from cheese. Foods 12:3937
8. Feichtinger M, Mayrhofer S, Kneifel W, Domig KJ (2016) Tetracycline resistance patterns of *Lactobacillus buchneri* group strains. J Food Prot 79:1741–1747
9. Výrostková J, Regecová I, Dudriková E, Marcinčák S, Vargová M, Kováčová M, Maľová J (2021) Antimicrobial resistance of *Enterococcus* sp. isolated from sheep and goat cheeses. Foods 10:1844
10. Výrostková J, Regecová I, Kováčová M, Marcinčák S, Dudriková E, Maľová J (2020) Antimicrobial resistance of *Lactobacillus johnsonii* and *Lactobacillus zeae* in raw milk. Processes 8:1627
11. Okonechnikov K, Golosova O, Fursov M (2012) UGENE team. Unipro UGENE: a unified bioinformatics toolkit. Bioinformatics. 28:1166–1167
12. Heuer H, Krsek M, Baker P, Smalla K, Wellington EM (1997) Analysis of actinomycete communities by specific amplification of genes encoding 16S rRNA and gel-electrophoretic separation in denaturing gradients. Appl Env Microbiol 63:3233–3241
13. Walker AW, Martin JC, Scott P, Parkhill J, Flint HJ, Scott KP (2015) 16S rRNA gene-based profiling of the human infant gut microbiota is strongly influenced by sample processing and PCR primer choice. Microbiome 3:26
14. Fuks G, Elgart M, Amir A, Zeisel A, Turnbaugh PJ, Soen Y, Shental N (2018) Combining 16S rRNA gene variable regions enables high-resolution microbial community profiling. Microbiome 6:17
15. BLAST: Basic Local Alignment Search Tool. https://blast.ncbi.nlm.nih.gov/Blast.cgi. Retrieved on 12 November 2025
16. Drancourt M, Bollet C, Carlioz A, Martelin R, Gayral JP, Raoult D (2000) 16S ribosomal DNA sequence analysis of a large collection of environmental and clinical unidentifiable bacterial isolates. J Clin Microbiol 38:3623–3630

Check for updates

Chapter 2

Generation of Custom Antibiotic-Resistant Variants of *Salmonella* Typhimurium by Genetic Engineering

Raúl Campillo, Ivo García-Penas, Natalia Merino, Daniel Berdejo, Rafael Pagán and Diego García-Gonzalo

Abstract

Antimicrobial resistance (AMR) remains a major public health concern affecting across the globe. Specifically, lots of efforts have been undertaken to counteract the acquisition of antimicrobial resistance genes (ARGs) in foodborne pathogens. Spontaneous mutations and/or horizontal gene transfer leading to acquisition of ARGs can inadvertently cross-protect microbes to other, unrelated microbial control agents used throughout the agri-food chain (e.g., food preservation methods, biocides, etc.). Therefore, what should we expect from a non-AMR foodborne pathogen acquiring ARGs? An effective approach to address this challenge is to study the phenotypic profile of customized AMR variants derived from parental foodborne strains with well-described genomes. In this chapter, we describe two methodologies used to transform *Salmonella* Typhimurium LT2 (SeT) with either extrachromosomal plasmids carrying ARGs (Protocol No. 1—Plasmid Transformation), or linear DNA fragments carrying a specific mutation for stable incorporation into the genome by substituting parental loci of ARGs (Protocol No. 2—Gene Editing). The application of any of these two protocols facilitates the construction of strains with defined genotypic and phenotypic profiles, aiding in the characterisation of their resistance to antibiotics and microbial control agents used in the agri-food chain. Both protocols can be easily adapted to a wide range of Gram-negative bacteria.

Keywords Antimicrobials, Antibiotic resistance, Electroporation, Plasmid transformation, Gene editing, Homologous recombination, Antimicrobial resistance genes, Bacteria, *Salmonella* Typhimurium, Sanger sequencing

1 Introduction

The rise of antimicrobial resistance (AMR) in pathogenic microorganisms poses a critical challenge to public health on a global scale [1]. The agri-food chain has emerged as a hotspot for emergence and dissemination of AMR bacteria [2], including *Salmonella* spp., a leading cause of foodborne outbreaks [3]. Its remarkable versatility and adaptability to harsh conditions [4] has turned *Salmonella* spp. into a major reservoir of antimicrobial resistancegenes (ARGs),

Marciane Magnani and Geany Targino de Souza Pedrosa (eds.), *Detection and Quantification of Bacteria, Yeast, Viruses, and Protozoan in Foods and Freshwater*, Methods and Protocols in Food Science, https://doi.org/10.1007/978-1-0716-5064-6_2,

which limits the efficacy of commonly used antibiotics, including those considered as last-resort alternatives [5].

Despite the apparent interconnection between the mechanisms driving AMR and protection to microbial control agents used throughout the agri-food chain (e.g., food preservation methods, biocides, etc.) [6–8], the impact of AMR on the efficacy of these agents remains unclear. To address this issue, an effective approach is to introduce ARGs into well-characterized bacterial strains, creating custom AMR variants that provide a clearer picture of how AMR in bacteria works and whether it can provide either cross-protection or cross-sensitization to food industry-related microbial control agents.

Reference strains, such as the widely used parental strain *Salmonella enterica* subsp. *enterica* serovar Typhimurium str. LT2 (SeT) [9], are ideal models for studying AMR due to their well-characterised genomes. By transforming SeT with ARGs (i.e., facilitating the uptake of DNA molecules through the cell wall from the external environment), we can generate derived customized variants to explore their effect on the efficacy of food industry-related microbial control agents via comparing their resistance with the parental strain. These custom AMR strains could additionally enhance our understanding of the molecular pathways hindering antibiotic action, which is crucial to pave the way towards novel antibiotic synergies and discovering new antimicrobial compounds [10, 11].

In this chapter, we describe two protocols to perform bacterial transformations with ARGs using either extrachromosomal plasmids (Protocol No. 1—Plasmid Transformation), or linear DNA fragments carrying a specific mutation for stable incorporation into the genome by substituting parental loci of ARGs (Protocol No. 2—Gene Editing) [12, 13]. Both protocols were successfully conducted in the past to deepen into the mechanisms of bacterial protection to heat treatments and essential oils [14, 15], firstly by transforming SeT with plasmid pKD46, and thereafter with a linear DNA fragment to introduce mutations in chromosomal ARGs (i.e., gene editing).

Our objective now is generating custom AMR variants of SeT with extrachromosomal plasmids (Protocol No. 1) or mutated chromosomal ARGs (Protocol No. 2). In this case, extrachromosomal transformation with ARGs is performed with plasmid pGDP4:*ermC* from the Antibiotic Resistance Platform [16] and the gene editing (i.e., integration of linear DNA carrying mutations into chromosomal ARGs) is conducted with a previously described 776 bp-long mutated sequence from the *ramR* gene.

2 Materials

As schematized in Fig. 1, extrachromosomal transformation with ARGs (Protocol No. 1) and chromosomal gene editing (Protocol No. 2) share the common need of rendering the bacteria competent

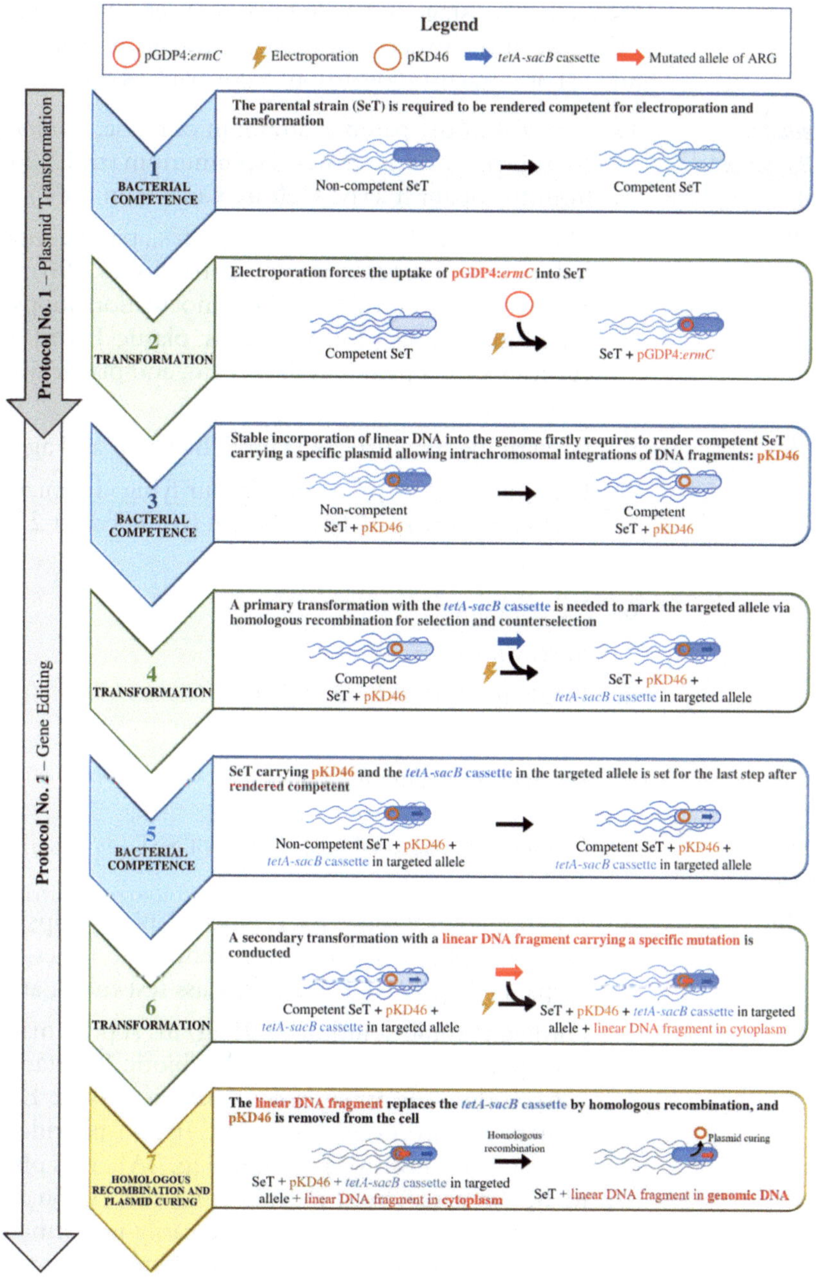

Fig. 1 Diagram depicting the steps to perform extrachromosomal transformations with an ARG-carrying plasmid (Protocol No. 1 – Plasmid Transformation), or intrachromosomal integration of linear DNA for stable incorporation in the bacterial genome (Protocol No. 2—Gene Editing). For illustrative purposes, note that Protocol No. 1 is schematized with pGDP4:*ermC*, but Protocol No. 2 requires transformation with pKD46

for electroporation and transformation. Competent bacteria are ideal for transformations due to the concurrence of several factors, such as increased membrane permeability, high uptake capacity of exogenous

DNA, etc. This stage is detailed below (Sects. 2.1 and 3.1), as well as the rest of the process either for extrachromosomal (Sects. 2.2 and 3.2) or intrachromosomal transformations (Sects. 2.3 and 3.3).

2.1 Generation of Competent Bacterial Cells for ARG Transformation

1. Cryovial of the parental strain: in our case, *Salmonella enterica* subsp. *enterica* serovar Typhimurium str. LT2 CECT 722 from the Spanish Type Culture Collection (SeT).
2. Microbiological tools: micropipettes, pipette tips, serological pipettes, pipette pump, petri dishes (90 mm), plastic 1.5 mL microcentrifuge tubes, inoculation loops, glass test tubes, glass flasks (250 mL), caps, plastic 1.5 mL semi-micro cuvettes with a path length of 1 cm, and plastic 50 mL centrifuge tubes.
3. Growth media: Luria–Bertani (LB) broth and agar.
4. Ultrapure water: prepared via purifying deionized water to attain a resistivity of 18.2 MΩ·cm (rounded) at 25°C.
5. Glycerol.
6. Ice container.
7. Vortex mixer.
8. Incubator with orbital shaker.
9. Spectrophotometer.
10. Refrigerated centrifuge.

2.2 Extrachromosomal Plasmid Transformation

1. Competent bacterial suspension of SeT (as described in Sect. 3.1).
2. Microbiological tools: micropipettes, pipette tips, petri dishes (90 mm), plastic 1.5 mL microcentrifuge tubes, inoculation loops, L-shaped cell spreaders, glass test tubes, and caps.
3. Purified plasmid solution (0.01 μg/μL approximately): in our case, pGDP4:*ermC* from the Antibiotic Resistance Platform [16], coding for *Staphylococcus aureus* enzyme ErmC, which confers erythromycin resistance. It also provides ampicillin resistance for positive selection (Fig. 2A). This plasmid needs to be extracted from a *Escherichia coli* DH5α, which contains pGDP4:*ermC*. This can be easily performed following the procedure of any commercial plasmid extraction kit. The strain mentioned above can be purchased on websites such as Addgene (https://www.addgene.org/; plasmid #112884) and comes in an agar stab from which colonies can be picked for culturing at 37°C.
4. Chilled sterile 0.2 cm gap electroporation cuvettes.
5. Growth media: Luria–Bertani (LB) broth and agar, prewarmed to 30°C.
6. Ampicillin sodium salt: to prepare a 100 mg/mL stock.

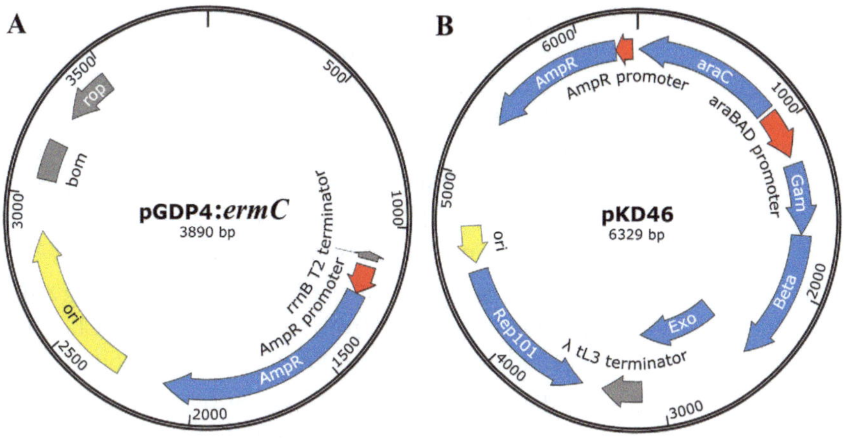

Fig. 2 Vector map of the main elements present in (**A**) pGDP4:*ermC*: *ori*: minimal origin of replication; *bom*: "basis of movility," element enabling plasmid transference between cells via conjugation; *rop*: "repressor of primer," encodes a protein regulating the number of plasmids in each cell; *rnbT2 terminator*: transcription termination sequence; AmpR promoter: regulates the expression of the ampicillin resistance protein (AmpR); AmpR: provides resistance to ampicillin for positive selection, and (**B**) pKD46: *ori*: minimal origin of replication; Rep101: replication protein for plasmid replication; *araC*: encodes AraC which positively regulates the expression of genes under the *araBAD* promoter in the presence of L-arabinose; *araBAD* promoter: regulates the expression of the λ Red recombinase genes (coding for Gam, Beta, Exo, and the λ tLr terminator); Exo, Beta, Gam: the λ Red recombinase proteins facilitating homologous recombination; *λ tL3 terminator*: transcription termination sequence; AmpR promoter: regulates the expression of the ampicillin resistance protein (AmpR); AmpR: provides resistance to ampicillin for positive selection. bp: base pairs

7. Electroporation system: in our case, Bio-Rad Gene Pulser Xcell™.
8. Phosphate-buffered saline (PBS).
9. Ice container.
10. Vortex mixer.
11. Incubator with orbital shaker.

2.3 Intrachromosomal Transformation for Gene Editing

1. Bacterial suspension transformed with pKD46 (as described in Sects. 3.1 and 3.2). Transformation with pKD46 allows chromosomal integrations [17] while providing ampicillin resistance for positive selection (Fig. 2B).
2. Microbiological tools: micropipettes, pipette tips, serological pipettes, pipette pump, petri dishes (90 mm), plastic 1.5 mL microcentrifuge tubes, inoculation loops, L-shaped cell spreaders, glass test tubes, glass flasks (250 mL), caps, plastic 1.5 mL semi-micro cuvettes with a path length of 1 cm, and plastic 50 mL centrifuge tubes.
3. Purified solution of a mutated allele of ARG (1 μg/μL approximately): in our case, a 776 bp-long DNA mutated

sequence from the *ramR* gene which confers AMR to SeT. This sequence was obtained from a SeT AMR variant with Primers 1 and 2 listed in Table 1. Alternative ARGs can be amplified from previously sequenced AMR strains by designing a pair of primers targeting the regions between –500 to –250 and +250 to +500 relative to the mutation in ARG (Fig. 3), providing a long enough amplicon to confirm intrachromosomal integration by Sanger sequencing.

4. Purified solution of *tetA-sacB* cassette (1 μg/μL approximately): amplified from *Escherichia coli* XTL298 [12] by high-fidelity PCR with Primers 3 and 4 from Table 1.
5. Purified solution of *tetA-sacB* cassette amplicon with homologous sequences for targeting (1 μg/μL approximately) (Fig. 4): the *tetA-sacB* cassette was used as the PCR template to produce an amplicon for targeting the dual cassette to the allele of choice (region of *ramR*), and induce the primary chromosomal integration. This PCR was performed using Phusion® High-Fidelity DNA Polymerase and specific chimeric primers with 50 extra base pairs (bp) homologous sequences located at the 3′-ends (Primers 3 and 4 from Table 1). The homologous sequences of this pair of primers are the 50 nucleotides after leaving a 50-nucleotide gap towards the mutation in ARG (Fig. 3).
6. TetA/SacB counter-selection medium: strictly follow the preparation instructions for this medium as described elsewhere [12]. Briefly, to prepare 350 mL dissolve 5.25 g of bacteriological agar, 1.4 g of tryptone, and 1.4 g of yeast extract in 140 mL of distilled water in a 500 mL flask. In a separate flask, dissolve 3.12 g of $NaH_2PO_4{\cdot}2(H_2O)$ and 2.8 g of NaCl in another 140 mL of distilled water. In a third flask, dissolve 21 g of sucrose in 35 mL of distilled water, and prepare additional 40 mL of distilled water in a final flask. Autoclave separately, mix the first three flasks in the 500-mL one and cool the mixture to 55°C. Then, add 175 μL from a 2000× stock solution of fusaric acid in ethanol (46 mg/mL) preserved at –20°C, 11.2 mL from a filter-sterilized (using a 0.22 μm pore size) solution of 25 mM $ZnCl_2$ in distilled water, and bring the volume to 350 mL with water from the fourth flask. Pour around 40 mL of the final molten mixture into petri dishes, let the surface slightly dry and store them in tightly closed plastic bags at 4°C. It is important not to excessively dry the plates to ensure they will keep hydrated during several days of incubation at 42°C. Under this temperature, the counter-selective effect of fusaric acid on *tetA* is very efficient, while permitting pKD46 curing (i.e., plasmid ejection from bacteria). Whether precipitates form, try again increasing the media volume in the plate and/or the humidity in the incubation

Table 1

Specific primers for PCR amplifications of the *ramR* target allele (Primers 1 and 2); and for targeting the *tetA-sacB* cassette to the desired *ramR* region of *Salmonella enterica* subsp. *enterica* serovar Typhimurium str. LT2 parental strain (Primers 3 and 4)

Primers	Sequences (5′ → 3′)
Primer 1 (Fw)	ATATCGTCGGCCCGTTGTTC
Primer 2 (Rv)	CCACTTGGAATACCCCGCAT
Primer 3 (Fw)	CACTCGACAATCGTGTCGATAACCTGAGCGGAAA TGGTCATCGTG CTCTC **TCCTAATTTTTGTTGACACTCTATC**
Primer 4 (Rv)	GGCGCGCGGGATGGTTCAGACCCCAACTGATGTA ACTGTTCCAGA TAAAA**ATCAAAGGGAAAACTGTCCATATGC**

Boldface in Primers 3 and 4 indicates *tetA-sacB* homologous regions and the rest of the sequence corresponds to homologous region for gene targeting. *Fw*: forward; *Rv*: reverse

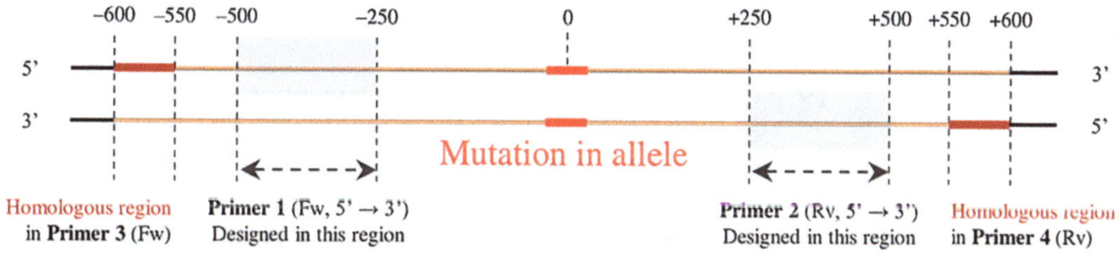

Fig. 3 Illustration schematizing the regions from which Primers 1 and 2 were obtained. Numbers indicate the relative positions to the mutated allele of ARG (bp). The homologous regions of Primers 3 and 4 for gene targeting are indicated in dark red. Fw: forward, Rv: reverse

environment. This could be originated by deposition of Zn^{+2} reacting with phosphate salts, which may affect the selective activity of fusaric acid.

7. Growth media: Luria–Bertani (LB) broth and agar, pre-warmed to the optimal growth temperature (for SeT, 37°C).
8. Ampicillin sodium salt: to prepare a 100 mg/mL stock.
9. Tetracycline hydrochloride: to prepare a 20 mg/mL stock.
10. L-Arabinose: to prepare a 20 mg/mL sterile stock after filter-sterilization with syringe using a 0.22 μm pore size.
11. Ultrapure water: prepared via purifying deionized water to attain a resistivity of 18.2 MΩ·cm (rounded) at 25°C.
12. Glycerol.
13. Ice container.
14. Vortex mixer.
15. Incubator with orbital shaker.

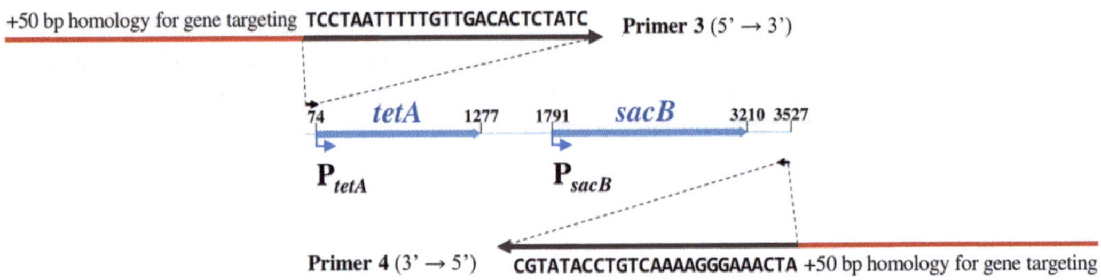

Fig. 4 Diagram of the *tetA*-*sacB* cassette and the primers used for adding homologous regions for targeting (Primers 3 and 4). Blue arrows indicate the transcription direction of both genes and its respective promoters. Numbers indicate the relative positions (bp). *tetA*: encodes TetA, which provides tetracycline resistance for positive selection and sensitivity to fusaric acid for negative selection; *sacB*: encodes SacB, which converts sucrose to levan, a toxic compound. The combination of the proteins TetA and SacB exerts independent toxic effects that enable removal by counter-selection

16. Spectrophotometer.
17. Refrigerated centrifuge.
18. Thermocycler.
19. Gel electrophoresis tools: electrophoresis grade agarose, electrophoresis gel casting tray, electrophoresis tank, power supply, TAE buffer, loading dye, DNA ladder (250–10,000 bp), UV-transilluminator, plastic combs for gel wells, micropipettes, pipette tips, SYBR Safe (DNA stainer), heat-resistant flask or beaker (for melting agarose), microwave or hot plate (for gel preparation), and parafilm or similar material for preparing loading dye mixtures.

3 Methods

3.1 Generation of Competent Bacterial Cells for ARG Transformation

Prior to the transformation with either a plasmid (Protocol No. 1) or linear DNA for gene editing (Protocol No. 2), bacteria are required to be rendered competent for electroporation and transformation. The steps listed below (1–11) are shared for both protocols.

1. From the cryovial of SeT, inoculate and streak over the surface of a LB agar plate with an inoculation loop to isolate individual colonies.
2. Incubate the LB agar plate for 24 h at 37°C (*see* **Note 1**).
3. Inoculate a single colony in a test tube containing 5 mL of LB broth (*see* **Note 2**).
4. Incubate the inoculated test tube overnight at 37°C and 140 rpm (revolutions per minute) until obtaining a stationary phase subculture (*see* **Note 3**).

5. Dilute the stationary phase subculture 100-fold in a flask containing 50 mL of LB broth (*see* **Note 4**).
6. Incubate the inoculated flask until obtaining a mid-exponential phase culture with an optical density of 0.4–0.6 measured at 600 nm (OD_{600}) using semi-micro cuvettes with a path length of 1 cm (*see* **Note 5**).
7. Chill the bacterial culture in ice and transfer the whole volume to a plastic 50 mL centrifuge tube.
8. Centrifuge the 50 mL tube for 20 min, 4000×*g* at 4°C, discard supernatant, and resuspend in 50 mL of ice-cold ultrapure water (*see* **Note 6**).
9. Repeat Step 8.
10. Repeat Step 8, but resuspending in 20 mL of ice-cold ultrapure water with 10% glycerol (v/v) (*see* **Note 7**).
11. Repeat Step 8, but resuspending in 1 mL of ice-cold ultrapure water with 10% glycerol (v/v).

3.2 Extrachromosomal Plasmid Transformation

1. Obtain a competent bacterial suspension (as described in Sect. 3.1).
2. Mix an aliquot of 40 μL from the competent bacterial suspension with 2 μL of the pGDP4:*ermC* solution in a chilled 1.5 mL microcentrifuge tube (*see* **Note 8**).
3. Transfer the whole volume to the electroporation cuvette and apply a single pulse of 12.5 kV/cm (2.5 kV, 200 Ω, 25 μF) (*see* **Note 9**).
4. Immediately add 1 mL of pre-warmed LB broth (37°C) and transfer the whole volume to a 1.5 mL microcentrifuge tube (*see* **Note 10**).
5. Incubate the bacteria for at least 1 h at 37°C.
6. Inoculate and spread on LB agar plates with 100 μg/mL of ampicillin to isolate individual colonies that incorporated pGDP4:*ermC* (*see* **Note 11**).
7. Incubate the LB agar plate for 24 h at 37°C.
8. Select several colonies and store them in cryovials (*see* **Note 12**).

3.3 Intrachromosomal Transformation for Gene Editing

To perform an intrachromosomal integration of linear DNA for gene editing (Protocol No. 2), it is required marking the targeted allele with the *tetA-sacB* cassette to properly conduct the process of selection and counterselection. For this primary double-crossover intrachromosomal integration, we firstly need to render competent SeT (as described in Sect. 3.1), which needs to have already been transformed with pKD46 [17]. Transformation with pKD46 can be performed as described with pGDP4:*ermC* in Sect. 3.2, but with few modifications (*see* **Note 10**). Also note that marking the targeted gene lead to its loss of function (*see* **Note 13**).

1. From the cryovial of SeT transformed with pKD46, inoculate and streak over the surface of a LB agar plate with 100 µg/mL of ampicillin using an inoculation loop to isolate individual colonies.
2. Incubate the LB agar plate for 24 h at 30°C (*see* **Note 1**).
3. Inoculate a single colony in a test tube containing 5 mL of LB broth with 100 µg/mL of ampicillin (*see* **Note 2**).
4. Incubate the inoculated test tube overnight at 30°C and 140 rpm until obtaining a stationary phase subculture (*see* **Note 3**).
5. Dilute the stationary phase subculture 100-fold in a flask containing 50 mL of LB broth with 100 µg/mL of ampicillin (*see* **Note 4**).
6. Add to the flask arabinose to a final concentration of 0.4 mg/mL after 2 h of incubation to induce the *araBAD* promoter expression.
7. Incubate the inoculated flask until obtaining a mid-exponential phase culture with an optical density of 0.4–0.6 measured at 600 nm (OD_{600}) with semi-micro cuvettes with a path length of 1 cm.
8. Perform Steps 7–11 from Sect. 3.1.
9. Mix an aliquot of 40 µL from the competent bacterial suspension with 2 µL from the solution of *tetA-sacB* cassette amplicon with homologous sequences for targeting in a chilled 1.5 mL microcentrifuge tube (*see* **Note 8**).
10. Transfer the whole volume to the electroporation cuvette and apply a single pulse of 12.5 kV/cm (2.5 kV, 200 Ω, 25 µF) (*see* **Note 9**).
11. Immediately add 1 mL of pre-warmed LB broth (30°C) with 100 µg/mL of ampicillin and transfer the whole volume to a 1.5 mL microcentrifuge tube.
12. Incubate the bacteria for at least 1 h at 30°C.
13. Inoculate and spread on LB agar plates with 20 µg/mL of tetracycline to isolate individual colonies that incorporated the *tetA-sacB* cassette, and 100 µg/mL of ampicillin to avoid pKD46 curing (*see* **Note 14**).
14. Incubate the plates for 24 h at 30°C.
15. Select several colonies and confirm the *tetA-sacB* cassette is incorporated into the targeted allele via PCR and electrophoresis (120 V for 30 min using a 1% w/v agarose gel prepared

with 5 μL of SYBR Safe and TAE buffer) with Primers 1 and 2 listed in Table 1. Verify that an approximate 3500-bp band is observed upon PCR and electrophoresis, confirming successful *tetA-sacB* integration into the targeted region. Store the confirmed colonies in cryovials.

Once the the *tetA-sacB* cassette is located into the target allele, it is necessary to repeat the competence for electroporation and transformation protocol, but this time with the *tetA-sacB*-marked strain. Thereupon, we can conduct the transformation with the mutated allele of ARG (secondary double-crossover intrachromosomal integration), as illustrated in Fig. 5.

16. Perform Steps 1–7 from this section, but adding 20 μg/mL of tetracycline to the ampicillin-containing LB agar plate from Step 1 (*see* **Note 14**). Optionally, subcultures and cultures of *tetA-sacB*-marked strain can be supplemented with additional 20 μg/mL of tetracycline to prevent spontaneous loss of the *tetA-sacB* cassette.
17. Mix an aliquot of 40 μL from the competent bacterial suspension with 2 μL from the solution of the mutated allele in a chilled 1.5 mL microcentrifuge tube (*see* **Note 8**).
18. Transfer the whole volume to the electroporation cuvette and apply a single pulse of 12.5 kV/cm (2.5 kV, 200 Ω, 25 μF) (*see* **Note 9**).
19. Immediately add 1 mL of pre-warmed LB broth (30°C) and transfer the whole volume to a 1.5 mL microcentrifuge tube.
20. Incubate the bacteria for at least 4 h at 30°C.
21. Inoculate and spread on pre-warmed (42°C) TetA/SacB counter-selection agar plates to isolate individual colonies.
22. Incubate the plates for 2 days at 42°C.
23. Select several colonies and serially re-streak them in LB agar, TetA/SacB counter-selection agar, and LB agar plates separately containing 100 μg/mL of ampicillin and 20 μg/mL of tetracycline (*see* **Note 15**).
24. Confirm the *tetA-sacB* cassette has been replaced in the colonies that lost sensitivity to fusaric acid and sucrose via high-fidelity PCR and electrophoresis (120 V for 30 min using a 1% w/v agarose gel prepared with 5 μL of SYBR Safe and TAE buffer) with Primers 1 and 2 listed in Table 1 (*see* **Note 15**). Store the confirmed ones in cryovials.
25. Submit the PCR product from Step 24 to Sanger sequencing to confirm the successful integration.

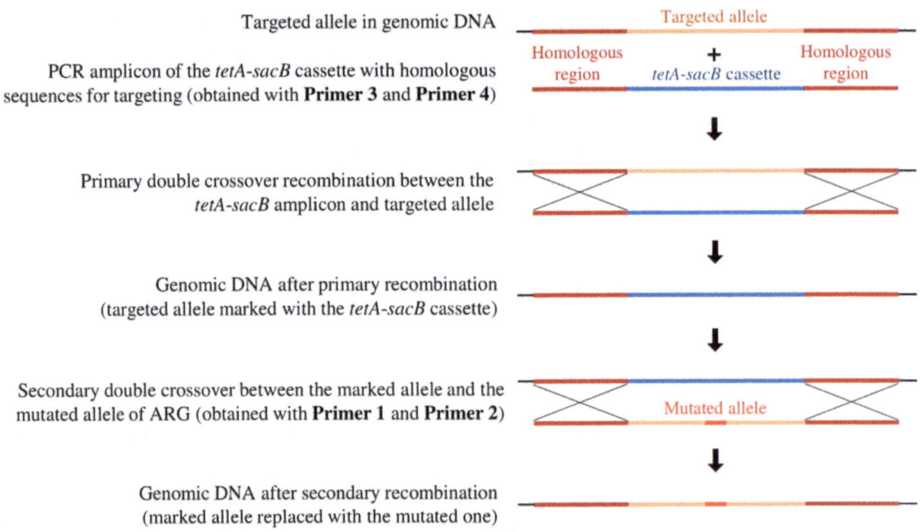

Fig. 5 Illustration of the double-crossover homologous recombinations performed for intrachromosomal transformation (gene editing). Primary chromosomal integration is performed to mark the targeted allele with the *tetA-sacB* cassette, and secondary chromosomal integration replaces it with the mutated allele of ARG

Once the ARG-transformed variants are obtained using Protocol No. 1 and/or Protocol No. 2, the impact of AMR on food industry-related microbial control agents can be properly assessed. This includes evaluating their inactivation rates under heat treatments or exposure to preservatives and comparing these with the parental strain. Additionally, their growth fitness at refrigeration temperatures, as well as tolerance to organic acids, sodium chloride, and other stressors, would provide valuable insights into cross-protections.

4 Notes

1. Incubation parameters should be adapted to the optimal growth conditions of the bacterial strain to study.
2. Verify homogeneity in size, color and shape among the colonies to avoid microbial contamination.
3. Incubation aiming to the stationary phase subculture should be adapted to the optimal growth conditions of the bacterial strain to study.
4. Ensure the bacteria from the subculture are homogeneously suspended before inoculating the flask by vortexing the test tube.
5. Under these conditions, mid-exponential growth phase of Gram-negative bacteria, as *Salmonella* spp., is usually linked

to OD_{600} values around 0.4–0.6 after 1.0–1.5 h of incubation in LB broth at 37°C. Whether incubated at 30°C (e.g., when containing pKD46 for Protocol No. 2—Gene Editing), SeT may take 3.5–4.0 h to reach mid-exponential growth phase.

6. To preserve the integrity of the bacterial membranes, ensure that the temperature of the bacterial suspension does not unnecessarily increase. We recommend placing the tube containing this suspension in ice between steps. We also suggest to perform the resuspensions via pipetting and never by vortex mixing, and to ensure that centrifugation does not exceed 4000 × *g* to prevent bacterial lysis.
7. An addition of glycerol is recommended for both maintaining membrane integrity during electroporation and avoiding arcing (i.e., uncontrolled discharge due to excessive voltage) by reducing the conductivity of the bacterial suspension. It also protects the bacterial membranes during freezing, allowing new electroporation attempts after storing the excess of competent bacterial suspension at −80°C.
8. The volumes can be adjusted for better yield. If doing so, ensure these changes are not excessively increasing the conductivity and significantly affecting the resistance (200 Ω), as well as the distribution of the electric field strength (12.5 kV/cm) throughout the whole volume.
9. To avoid arcing, remove any water droplets that may remain on the outside surface of the electroporation cuvette and do not touch the aluminum electrodes. Ensure that no bubbles are formed in the mixture before electroporation. Both overly concentrated or diluted bacterial suspensions may lead to arcing. Whether arcs occur, try diluting the competent bacterial suspension to increase conductivity, or concentrate it to prevent uncontrolled discharge. Under our conditions, we have regularly electroporated suspensions with a concentration of approximately 5×10^9 CFU/mL (colony-forming units). Cracked or aged cuvettes may also cause arcing.
10. Note that plasmids may carry thermosensitive replication systems, as Rep101 in pKD46 (Fig. 2B), which are recommended to be replicated at temperatures below 32°C (most commonly at 30°C). Check if the plasmid you are working with can replicate at higher temperatures, meaning shorter incubation times.
11. Ensure your parental strain is susceptible to 100 μg/mL of ampicillin. Always supplement media with ampicillin after autoclaving, adding it once the media has cooled to approximately 50°C. The plate must be homogeneously supplemented with ampicillin (100 μg/mL) to prevent plasmid curing during bacterial growth. It might be necessary to serially dilute in PBS to properly isolate individual colonies.

12. To avoid selection of spontaneous resistant variants, confirm successful transformation via PCR and electrophoresis (120 V for 30 min using a 1% w/v agarose gel prepared with 5 μL of SYBR Safe and TAE buffer), or restriction digest. We recommend culturing the pGDP4:*ermC*-containing strains in presence of 100 μg/mL of ampicillin to prevent plasmid curing.
13. The integration of the *tetA-sacB* cassette knocks-out the targeted gene. Before marking, ensure the targeted gene is not essential for bacterial viability. If this is your case, aim to integrate the cassette immediately after the open reading frame of the targeted gene, or before the –35/–10 promoter sites.
14. Ensure your parental strain is susceptible to 20 μg/mL of tetracycline. Always supplement media with tetracycline after autoclaving, adding it once the media has cooled to approximately 50°C. The tetracycline stock and agar plates must be stored in lightproof containers and its exposure to light while manipulation controlled.
15. Re-streaking the colonies that grew in the TetA/SacB counter-selection agar after the transformation will discriminate the ones that ejected pKD46 and replaced the *tetA-sacB* cassette due to a successful intrachromosomal integration with the mutated allele of ARG. We don't recommend performing Protocol No. 2 to transform with linear DNA than 1000 bp, as it may negatively affect recombination efficiency. If this is your case, increasing the size of the homology arms in the chimeric primers (Primers 3 and 4) may help.

References

1. European Centre for Disease Prevention and Control and World Health Organization (2023) Antimicrobial resistance surveillance in Europe 2023–2021 data. EFSA and ECDC, Stockholm
2. Kumar SB, Arnipalli SR, Ziouzenkova O (2020) Antibiotics in food chain: the consequences for antibiotic resistance. Antibiotics 9:688
3. European Food Safety Authority (EFSA) and European Centre for Disease Prevention and Control (ECDC) (2023) The European Union One Health 2022 zoonoses report. EFSA J 21:e8442
4. Pradhan D, Negi VD (2019) Stress-induced adaptations in *Salmonella*: a ground for shaping its pathogenesis. Microbiol Res 229:126311
5. World Health Organization (2019) Critically important antimicrobials for human medicine, 6th revision. WHO, Geneva
6. Giacometti F, Shirzad-Aski H, Ferreira S (2021) Antimicrobials and food-related stresses as selective factors for antibiotic resistance along the farm to fork continuum. Antibiotics 10:671
7. Liao X, Ma Y, Daliri EBM et al (2020) Interplay of antibiotic resistance and food-associated stress tolerance in foodborne pathogens. Trends Food Sci Technol 95:97–106
8. Berdejo D, Gayán E, Pagán E et al (2022) Carvacrol selective pressure allows the occurrence of genetic resistant variants of *Listeria monocytogenes* EGD-e. Foods 11:3282
9. McClelland M, Sanderson KE, Spieth J et al (2001) Complete genome sequence of *Salmonella enterica* serovar Typhimurium LT2. Nature 413:852–856
10. Brown ED, Wright GD (2016) Antibacterial drug discovery in the resistance era. Nature 529:336–343
11. Hoffman PS (2020) Antibacterial discovery: 21st century challenges. Antibiotics 9:213
12. Li XT, Thomason LC, Sawitzke JA et al (2013) Positive and negative selection using

the *tetA-sacB* cassette: recombineering and P1 transduction in *Escherichia coli*. Nucleic Acids Res 41:e204
13. O'Callaghan D, Charbit A (1990) High efficiency transformation of *Salmonella* Typhimurium and *Salmonella* Typhi by electroporation. Mol Gen Genet 223:156–158
14. Pagan E, Merino N, Berdejo D et al (2024) Adaptive evolution of *Salmonella* Typhimurium LT2 exposed to carvacrol lacks a uniform pattern. Appl Microbiol Biotechnol 108:38
15. Berdejo D, Pagán E, Merino N et al (2021) *Salmonella enterica* serovar Typhimurium genetic variants isolated after lethal treatment with *Thymbra capitata* essential oil (TCO) showed increased resistance to TCO in milk. Int J Food Microbiol 360:109443
16. Cox G, Sieron A, King AM et al (2017) A common platform for antibiotic dereplication and adjuvant discovery. Cell Chem Biol 24:98–109
17. Datsenko KA, Wanner BL (2000) One-step inactivation of chromosomal genes in *Escherichia coli* K-12 using PCR products. Proc Natl Acad Sci 97:6640–6645

Chapter 3

Adaptive Evolution Assays Using Increasing Concentration of Antibiotics to Isolate Resistant Bacterial Variants and Determination of Genetic Changes via Whole Genome Sequencing

Ivo García-Penas, Raúl Campillo, Daniel Berdejo, Laura Espina, Diego García-Gonzalo and Rafael Pagán

Abstract

The widespread misuse and overuse of antibiotics in the agrifood sector have significantly contributed to the emergence of antibiotic-resistant bacteria, which pose a serious public health threat. Furthermore, because antimicrobials often share mechanisms of action, there is an increased risk of developing cross-tolerance. These phenomena underscore the need for an in-depth study of the molecular mechanisms behind resistance. In this context, the deliberate generation of antibiotic-resistant variants (RVs) provides a valuable tool for investigating the mechanisms of resistance acquisition and maintenance. This chapter outlines a step-by-step protocol for obtaining RVs through exposure to progressively increasing doses of antibiotics. Additionally, it presents a methodology for the detection and annotation of mutations in the RVs. By mapping and analyzing these mutations, researchers can better predict the role of specific genetic changes in the development of different forms of antibiotic resistance, offering insights into potential strategies for combating this global challenge.

Keywords Antimicrobials, Antibiotic resistance, Adaptive laboratory evolution, Whole genome sequencing, Resistant variant, Variant calling, Variant annotation, Bacteria, *Salmonella* Typhimurium

1 Introduction

The overuse and misuse of antibiotics over the past several decades have led to the emergence of various bacterial resistance phenomena. Consequently, antibiotics are becoming increasingly ineffective, with some microorganisms developing partial or full resistance, a condition known as antimicrobial resistance (AMR). AMR complicates the treatment of infections, leading to longer hospital stays, increased treatment costs, and higher mortality rates. The economic burden of these infections is substantial, with estimated costs of approximately

Marciane Magnani and Geany Targino de Souza Pedrosa (eds.), *Detection and Quantification of Bacteria, Yeast, Viruses, and Protozoan in Foods and Freshwater*, Methods and Protocols in Food Science, https://doi.org/10.1007/978-1-0716-5064-6_3,

€9 trillion annually in the European Union and €55 trillion annually in the United States [1].

Among other pathways, these resistances can arise from genetic mutations that occur randomly during the replication of genetic material, which might be induced by antibiotic exposure [2, 3]. In this context, cells with genetic mutations that confer resistance to certain antibiotics coexist in the bacterial environment, as well as sensitive cells without such mutations. When the bacterial ecosystem is exposed to the presence of antibiotics, a selective pressure is created that allows the proliferation of cells that possess resistance over those that lack it. This results in a bacterial population with genetic material that encodes resistance characteristics. This phenomenon is a natural process, although the misuse of drugs contributes to its acceleration [4].

There is a wide variety of mechanisms by which micro-organisms can inhibit or face the antimicrobial activity of certain molecules. Traditionally these mechanisms have been classified into three groups: (1) changes in membrane permeability including increased expression of efflux pumps, (2) modification or destruction of the antimicrobial compound, or (3) modification of target structures. Nevertheless, acquired AMR can also involve non-canonical mechanisms. Recent studies on *Escherichia coli* indicate that mutations in metabolic genes are often underdiagnosed despite their significant role in antibiotic lethality [5, 6]. The unspecific nature of the resistance mechanisms leads to the possible emergence of cross-tolerance phenomena. Cross-tolerance occurs when resistance to antibiotics indirectly affects resistance to other antimicrobial agents such as food preservatives, biocides, and has been suggested to happen in processes used in the food industry for microbial control, like heat treatments [7–9].

Both direct and indirect resistance-related mutations can imply either an overexpression or repression of the functionality of the gene. In this framework, adaptive laboratory evolutions (ALE) have been devised to force the appearance of these mutations. Originally, these consisted in cyclic exposure of the cell population to sub-inhibitory doses of a specific antimicrobial [10], but current studies implement an increasing concentration gradient, a more aggressive technique that can yield more resistant bacterial variants (RVs) while rehearsing fewer exposure cycles [6, 11]. Whole genome sequencing (WGS) is generally carried out after ALE to perform downstream analyses, which can include genotypic-phenotypic mapping to identify mutations involved in AMR and other forms of resistance [12, 13]. Bioinformatic analysis of variant calling, leveraging comparative genomics, is crucial to compare the reference genome sequence with that of the genetic variant to identify mutations that may be related to antibiotic resistance in microorganisms. This analysis can detect specific changes in the genetic sequence that alter the structure of key proteins or activate defense mechanisms, reducing the effectiveness of antimicrobial treatments. By knowing these mutations, the genetic processes behind resistance might be better understood. Also, predicting the response of different strains to specific antibiotics can contribute to the design of more effective therapies [14, 15]. This approach not only improves the diagnosis and treatment of resistant infections, but also is vital in epidemiological surveillance to control the spread of resistant pathogens, addressing one of the most pressing challenges in public health worldwide.

2 Materials

2.1 Minimum Inhibitory Concentration (MIC) and Adaptive Laboratory Evolution (ALE)

1. Cryovial of the bacterial strain to study.
2. Culturing tools: micropipettes, pipette tips, petri dishes, L-shaped spreaders.
3. Growth media: cation-adjusted Mueller-Hinton broth (CAMHB); tryptic soya broth and tryptic soya agar supplemented with 0.6% yeast extract (TSBYE or TSAYE).
4. Antibiotics to study.
5. Phosphate-buffered saline.
6. 96-well plates, glass test tubes, glass flasks (250 mL) and caps.
7. Vortex.
8. Incubator with orbital shaker.
9. Spectrophotometer plate reader.
10. Cryovials.

2.2 DNA Extraction, Whole Genome Sequencing (WGS) and Post-Sequencing Analysis of Each Evolved Strain for the Determination of Antibiotic Resistance Genes and of Biosynthetic Gene Clusters

2.2.1 DNA Extraction

1. Working bacterial culture of the parental and evolved strains.
2. DNA extraction kit (e.g., DNeasy Blood and Tissue Kits for DNA Isolation, QIAGEN, Hilden, Germany).
3. Culturing tools: micropipettes and pipette tips.
4. Microcentrifuge.

5. Microtubes.
6. Spectrophotometer for DNA quantification (e.g., Nanodrop 2000, Thermo Scientific, Massachusetts, EE.UU.).

2.2.2 Whole Genome Sequencing (WGS)

1. Illumina-based WGS service.

2.2.3 Post-Sequencing Analysis for the Detection of Bacterial Variants

1. Paired-end raw FASTQ files with short reads generated from the submitted DNA.
2. Sequence FASTA file and annotations GFF3 file of the reference strain.
3. Bioinformatics tools and their dependencies accessed on a local Linux server, using the European Galaxy Server [16], or via their standalone GUI application.
   3.1 Trim Galore! v0.6.7 [17].
   3.2 BWA-MEM v0.7.18 [18].
   3.3 SAMtools v2.0.5 [19].
   3.4 BCFtools v1.15.1 [20].
   3.5 SnpEff v4.3 [21].
4. Integrative genomics viewer [22].

3 Methods

3.1 Minimum Inhibitory Concentration (MIC) Determination

Prior to the ALE protocol, a working bacterial culture of the parental strain must be obtained from which ALEs will be triggered. Also, the minimum inhibitory concentration (MIC) for each tested antibiotic must be determined according to the Clinical Laboratoy Standards Institute (CLSI) specifications [23]. MIC is defined as the lowest concentration of antibiotic capable of inhibiting bacterial growth. This value will be used to establish the starting concentration of the ALE protocol.

Firstly, to obtain a working bacterial culture:

1. Inoculate from a cryovial of the parental strain and streak over TSAYE plates with an inoculation loop to obtain individual colonies.
2. Incubate at 37°C/24 h.
3. Inoculate a single colony in 5 mL of TSBYE in a test tube.
4. Incubate overnight at 37°C.

After this step, MIC values must be determined for each antibiotic as follows:

1. In a microwell plate distribute 200 μL of CAMHB with increasing concentration of the antibiotic in a 2× ratio.
2. Inoculate the plates with an equivalent of 0.5 McFarland standard, in this case 10^5 CFU/mL (colony forming units/mL).
3. A positive control inoculated with 10^5 CFU/mL without the tested antibiotic must be included, as well as a negative control without inoculum and with the tested antibiotic.
4. Incubate 37°C/20 h at 140 rpm (*see* **Note 1**)
5. Measure turbidity at optical density 595 (DO_{595}). Wells with a DO_{595} reduction of over 90% compared to the parental strain were considered to have inhibited growth due to antibiotic action [24].
6. MIC is established as the lowest concentration of antibiotic capable of inhibiting bacterial growth under the test conditions.

3.2 Adaptive Laboratory Evolution (ALE) with Increasing Concentrations of Antibiotic

Once the MIC value is obtained it can be used as reference for the first antibiotic exposure when carrying out the ALE protocol using increasing concentration of antibiotics. ALE started at 0.085× MIC, increasing the antibiotic concentration by 85% daily, thus reaching MIC on the 5th day of exposure and 40× MIC at 11th cycle of exposure (*see* **Note 2**). A schematic of the ALE can be found in Fig. 1.

This protocol involves applying increasing stress to a bacterial population across a steep concentration gradient, heightening the chances of selecting for characteristic RVs [25]. The objective is to favor selection of mutations that enhance bacteria's growth fitness in the presence of an antimicrobial agent. If beneficial mutations arise, the selective pressure of the antimicrobial agent will favor the proliferation of these RVs over the parental strain and other genetic variants that lack resistance-related mutations. This approach enables the isolation of bacterial RVs to the chosen antimicrobial compound (Fig. 2).

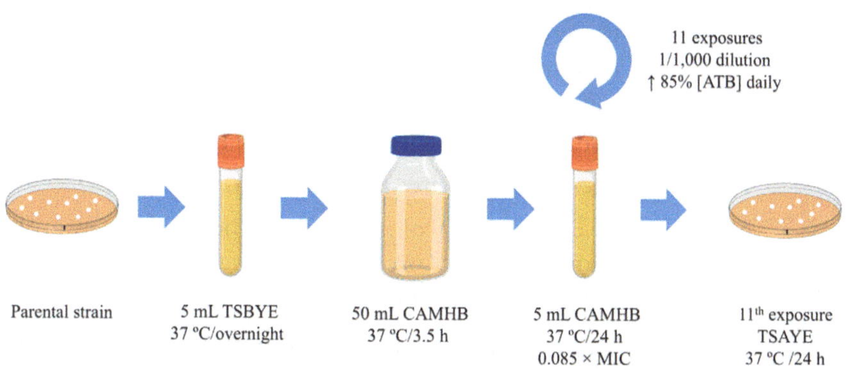

Fig. 1 Adaptive laboratory evolution with increasing concentrations of antibiotic

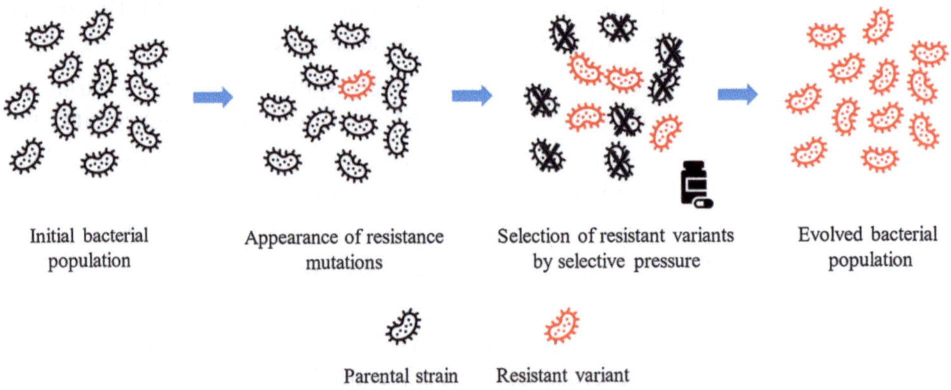

Fig. 2 Selection of resistant variants by cyclic exposure to increasing concentrations of antibiotic

To carry out the ALE, take the following steps after obtaining a working bacterial culture:

1. Dilute working bacterial culture of the parental strain 1:1000 into 50 mL of CAMHB and incubate 3.5–4 h at 37°C and 140 rpm until exponential phase starts.
2. Then, inoculate 5 mL CAMHB in test tubes at 10^6 CFU/mL with 0.085× MIC of the antibiotic.
3. Incubate at 37°C/24 h and 140 rpm to reach stationary phase.
4. From the grown tubes, inoculate 5 mL CAMHB test tubes with 1:1000 dilution in the presence of 1.85× the antibiotic concentration used on the previous day (*see* **Note 3**).
5. Incubate at 37°C/24 h and 140 rpm to reach stationary phase.
6. Return to Step 4. After 11 cycles of exposure, 40× MIC for the tested antibiotic will be reached in the test tubes, which is established as the endpoint of the ALE. Following the 5th cycle of exposure, in which the antibiotic concentration in the tube equals the MIC for the antibiotic, cryovials with 20% glycerol (v/v) must be stored at −80°C for each strain. Then, continue to step 7 (*see* **Note 4**).
7. Dilute the stored cryovials of each exposure in PBS, inoculate and spread on agar plates to obtain individual colonies.
8. Incubate the plates at 37°C/24 h.
9. Select several colonies from the highest exposure which allowed growth on the agar plates. Grow at 37°C/24 h and store separately in cryovials at −80°C.

In addition to the ALE, domestication control must be obtained via bacterial exposure in CAMHB without antibiotic. This will allow for the identification of mutations caused by adaptation to the growth medium, not the antibiotic.

Once the ALE is finished and evolved variants are obtained, it is recommended to assess their direct resistance to the tested antibiotic by determining their MIC value (as described in Sect. 3.1).

3.3 DNA Extraction, Whole Genome Sequencing (WGS) and Post-Sequencing Analysis

3.3.1 DNA Extraction

A pure culture of the parental and evolved strains must undergo DNA extraction with an adequate bacterial DNA extraction kit, following its instructions. The DNA concentration should be measured using a spectrophotometer, such as the Nanodrop 2000 (Thermo Scientific), to confirm a correct extraction.

3.3.2 Whole Genome Sequencing (WGS)

A sequencing provider can perform DNA quality checking, library preparation and sequencing on an Illumina platform, delivering high-quality sequencing reads with appropriate Phred scores.

3.3.3 Post-Sequencing Analysis for the Detection of Bacterial Variants

Data Pre-Processing

The next step aims to identify specific mutations in the bacterial genome. For this, the paired-end raw FASTQ files resulting from WGS are subjected to variant calling (i.e., identification of mutations in the bacterial genome) and variant annotation (i.e., addition of meaningful biological information to identified genetic variants). The sequence FASTA and the annotation GFF3 file of the reference strain used in the study must be downloaded (e.g., from NCBI's RefSeq Nucleotide Archive [26]). All bioinformatics analysis can be performed using Galaxy's European servers if access to a local server with the appropriate tools is not available. Galaxy is an online scientific analysis platform utilized by scientists worldwide for analyzing large-scale biomedical datasets in fields such as genomics, proteomics, metabolomics, and imaging [16]. The whole process is described below. All tools listed below can be accessed with Galaxy's inbuilt search engine.

1. Sequencing the parental and the evolved strains: WGS of the parental and the evolved strains must be performed. This can be performed by specialized companies, providing an output of paired-end raw FASTQ files (*see* **Notes 5 and 6**).
2. Next, the raw FASTQ files must be trimmed using Trim Galore! (based on Cutadapt [27]) Its main purpose is to clean

up raw sequencing reads by removing low-quality bases and adapter sequences, which are often added during library preparation. Inside the tool, both forward (R1) and reverse (R2) FASTQ files must be introduced in the tool to be processed in paired-end mode [17] (*see* **Note 7**). Trim Galore! v0.6.7 incorporates FastQC v0.74 [28] into its processing streamline, so it automatically performs quality control checks both before and after adapter trimming. This integration overrides the need to run FastQC as a separate step to check the quality of the trimmed reads. An acceptable Phred score is set at a value of 20, meaning there is 1 in 100 chance that the read base is incorrect. If another trimming tool is used (e.g., Cutadapt or Trimomatic) FastQC must be used to determine the quality of the reads.

Variant Calling

1. Read mapping and alignment using BWA-MEM (Burrows-Wheeler Aligner–Maximal Exact Matches): this tool is used to align sequencing reads to a reference genome. To this end, both trimmed reads forward (R1) and reverse (R2) must be submitted along with the reference genome (FASTA file) of the studied bacteria.
2. Use samtools-view to transform the SAM file obtained in the previous step to BAM format. Then, samtools-sort is used to reorder the reads in the BAM file according to their genomic coordinates, which is essential for various downstream analyses, including variant calling and visualization [20] (*see* **Note 8**).
3. Then, insert both the reference genome in FASTA format and the BAM file with aligned sequencing data into bcftools mpileup. This tool works by taking the aligned DNA sequencing data and analyzing each position in the reference genome to identify how many reads cover that position and what variants (such as SNPs or indels) may be present [19]. When bcftools mpileup is used with a reference FASTA file and an alignment BAM file, it automatically detects and uses the .fai and .bai index files if they are present in the same directory as their corresponding FASTA and BAM files.
4. Finally, the variant calling tool bcftools call is used to detect and list variations in specific positions of the genome. This tool reads the mpileup data obtained in the previous step and identifies differences between the reference genome and the variant genome, producing a detailed file listing the variants in bcf or vcf format (*see* **Note 9**).

Variant Annotation

This concludes the variant calling process. To further investigate the impact of these mutations, an annotation process can be performed. The annotation is crucial, as it predicts the biological effects of each variant. It helps to interpret the effect each variant will have on the genes, proteins and other regulatory elements, as well as their impact on resistance mechanisms.

1. First, a database for the studied bacteria must be built. To this end, SnpEff build will be used using the FASTA reference and GFF3 annotation file of the reference strain.
2. Once the database is constructed, use SnpEff Eff to annotate the variants. This will provide important information, including specific amino acid changes, the position within the affected gene, and the impact on protein translation (*see* **Note 10**).

Data Exploration

Next, an exploration of the data is required for gaining an initial understanding of the obtained information. By using a spreadsheet application, the data can be uploaded, filtered based on quality, and the results can be explored. During this stage, it is essential to focus on key aspects such as the variant effect, impact level, and predicted gene associated with each variant.

It is recommended to check the detected mutations in the Integrative Genomics Viewer (IGV) [22] to confirm their presence in the genome alignment. This check is important, as some variant calls may be due to errors in the alignment process or in the quality of the sequencing. By visualizing mutations in IGV, it is possible to directly assess whether they are supported by enough high-quality reads, which helps to ensure that the mutations identified are true and not technical artefacts (*see* **Note 11**).

This analysis must also be performed on the domestication control, thus mutations due to the exposure to growth medium can be discarded.

After confirming the mutations, the relevant data from the variant calling process can be organized and displayed in a table format, similar to Table 1, to present all the key information in a clear and structured manner. To help with the visualization and interpretation of the results of variant calling and annotation, graphical representations such as Venn diagrams and Voronoi maps can be extremely useful (Fig. 3). Venn diagrams allow to quickly compare the overlap of variants between different samples or populations, highlighting shared and unique variants. This is particularly helpful when comparing the genetic differences across multiple groups or identifying common mutations linked to other forms of tolerance (Fig. 3). On the other hand, Voronoi maps provide a more spatial approach by visually clustering variants based on their genomic locations or functional impact, offering insights into the distribution of mutations across genes or regions of interest (Fig. 4). Voronoi maps can be created using tools such as TreeMap (Macrofocus, Zürich, Switzerland) or with free packages compatible with R, such as deldir [29]. These types of visualizations simplify complex data, making it easier to identify patterns, relationships, and outliers, ultimately enhancing the interpretability of genomic studies.

Table 1

Examples of mutations occurring in *Salmonella enterica* subsp. *enterica* serovar Typhimurium str. LT2 evolved in the presence of increasing concentrations of a specific antibiotic

Genome position	Type of mutation	Variant effect	Change	Impact	Locus tag	Gene	Description
100,900	SNV	Missense	c.62C>T → p. Pro21Leu	Moderate	STM0087	*folA*	Dihydrofolate reductase
3,162,445	Frameshift insertion	Frameshift	c.259dupG → p. Asp87fs	High	STM3004	*ygdP*	RNA pyrophosphohydrolase
4,504,754	SNV	Missense	c.359T>C → p. Ile120Thr	Moderate	STM4266	*soxR*	Redox-sensitive transcriptional activator

SNV, single nucleotide variation; *dup*, duplication; *fs*, frameshift

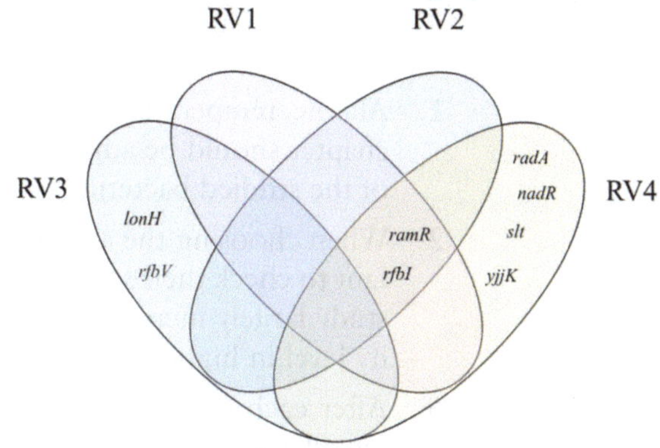

Fig. 3 Example of representation of gene confluence in four theoretical resistant variants (RV 1–4) by Venn diagrams

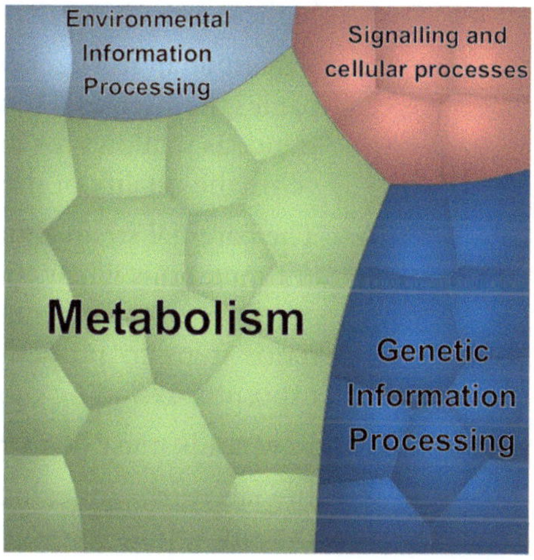

Fig. 4 Example of representation according to functional category in Voronoi map

Several studies have established a cross-resistance between antibiotics and other stresses, such as carvacrol and heat [9, 30]. So, submitting the RVs to different lethal treatments and comparing their tolerance patterns can yield essential insight on how each mutation and gene are implied in the cell's response to stress. These results will provide insights into the cellular responses and resistance mechanisms against both the antibiotic and other stresses. This information could aid in optimizing the treatments used in the food industry.

4 Notes

1. All the temperatures and times of incubation used in this chapter should be adjusted to the optimal growth conditions of the studied bacteria.
2. When choosing the antibiotics for the ALE assays it is important to check their solubility, as the concentrations used in the study largely increase and can potentially surpass the solubility level in higher exposures.
3. After each exposure, bacteria should be diluted in PBS and inoculated on TSAYE plates to ensure the absence of contamination by checking homogeneity in the morphology of the grown colonies.
4. Depending on the antibiotic, lineages may not reach the 11th exposure. In this case, bacteria should be isolated from the highest exposition that allowed growth.
5. Variant analysis is generally provided by the sequencing companies themselves. However, it is still a valuable tool to understand, as it allows for independent analysis to cross-check and verify the quality of the analysis provided by these companies.
6. The parental strain should always be sequenced, as it will present mutations when compared to the reference sequence. The sequence of the parental strain can be used as reference in posterior steps to properly identify mutations due to antibiotic exposure, excluding those already present in the parental strain.
7. Raw reads must be introduced as Paired-end within the tool.
8. In Galaxy, the indexing process is integrated into these tools. If using these tools outside the Galaxy environment, the genome must be indexed using samtools index (which creates a.bai file), and the reference FASTA file must be indexed using samtools faidx (which creates a.fai file).
9. It is essential to activate the option Output variant sites only in the output section of the tool. Otherwise, a file listing each position regardless of having a variation will be generated, increasing the size of the file and complicating the analysis process.
10. During the annotation process there can be problems with the database depending on the version of the tool used, as the commands may conflict with the content of the annotation file. Using older versions of the tool can help solve this issue.
11. This tool can also be used to check that the alignment process was successful by introducing the.bam file obtained with samtools sort.

References

1. Dadgostar P (2019) Antimicrobial resistance: implications and costs. Infect Drug Resist 12:3903–3910
2. Revitt-Mills SA, Robinson A (2020) Antibiotic-induced mutagenesis: under the microscope. Front Microbiol 11:585175
3. Rosenblatt-Farrell N (2009) The landscape of antibiotic resistance. Environ Health Perspect 117:A244–A250
4. Davies J, Davies D (2010) Origins and evolution of antibiotic resistance. Microbiol Mol Biol Rev 74:417–433
5. Munita JM, Arias CA (2016) Mechanisms of antibiotic resistance. Microbiol Spectr 4(2): 481–511
6. Lopatkin AJ, Bening SC, Manson AL, Stokes JM, Kohanski MA, Badran AH, Earl AM, Cheney NJ, Yang JH, Collins JJ (2021) Clinically relevant mutations in core metabolic genes confer antibiotic resistance. Science 371(6531):eaba0862
7. Walsh SE, Maillard JY, Russell AD, Catrenich CE, Charbonneau DL, Bartolo RG (2003) Development of bacterial resistance to several biocides and effects on antibiotic susceptibility. J Hosp Infect 55:98–107
8. Walsh C, Duffy G, Sheridan JJ, Fanning S, Blair IS, McDowell DA (2005) Thermal resistance of antibiotic-resistant and antibiotic-sensitive *Salmonella* spp. on chicken meat. J Food Saf 25:288–302
9. Pagan E, Merino N, Berdejo D, Campillo R, Gayan E, García-Gonzalo D, Pagan R (2024) Adaptive evolution of *Salmonella* Typhimurium LT2 exposed to carvacrol lacks a uniform pattern. Appl Microbiol Biotechnol 108:38
10. Andersson DI, Hughes D (2014) Microbiological effects of sublethal levels of antibiotics. Nat Rev Microbiol 12:465–478
11. Vinchhi R, Jena C, Matange N (2023) Adaptive laboratory evolution of antimicrobial resistance in bacteria for genetic and phenotypic analyses. STAR Protocols 4:102005
12. Patel V, Matange N (2021) Adaptation and compensation in a bacterial gene regulatory network evolving under antibiotic selection. eLife 10:e70931
13. Wistrand-Yuen E, Knopp M, Hjort K, Koskiniemi S, Berg OG, Andersson DI (2018) Evolution of high-level resistance during low-level antibiotic exposure. Nat Commun 9: 1599
14. van Belkum A, Dunne WM (2020) Next-Generation antimicrobial susceptibility testing. J Clin Microbiol 51:2018–2024
15. Zankari E, Hasman H, Kaas RS, Seyfarth AM, Agersø Y, Lund O, Larsen MV, Aarestrup FM (2013) Genotyping using whole-genome sequencing is a realistic alternative to surveillance based on phenotypic antimicrobial susceptibility testing. J Antimicrob Chemother 68:771–777
16. The Galaxy C (2024) The Galaxy platform for accessible, reproducible, and collaborative data analyses: 2024 update. Nucleic Acids Res 52(W1):W83–W94
17. Krueger F, James F, Ewels P, Afyounian E, Weinstein M, Schuster-Boeckler B, Hulselmans G, Sclamons (2023) FelixKrueger/TrimGalore: v0.6.10 - add default decompression path. Zenodo
18. Li H, Durbin R (2009) Fast and accurate short read alignment with Burrows–Wheeler transform. Bioinformatics 25:1754–1760
19. Li H, Handsaker B, Wysoker A, Fennell T, Ruan J, Homer N, Marth G, Abecasis G, Durbin R, Genome Project Data Processing S (2009) The sequence alignment/map format and SAMtools. Bioinformatics 25:2078–2079
20. Danecek P, Bonfield JK, Liddle J, Marshall J, Ohan V, Pollard MO, Whitwham A, Keane T, McCarthy SA, Davies RM, Li H (2021) Twelve years of SAMtools and BCFtools. GigaScience 10: giab008
21. Cingolani P, Platts A, Wang LL, Coon M, Nguyen T, Wang L, Land SJ, Lu X, Ruden DM (2012) A program for annotating and predicting the effects of single nucleotide polymorphisms, SnpEff. Fly 6:80–92
22. Robinson JT, Thorvaldsdóttir H, Winckler W, Guttman M, Lander ES, Getz G, Mesirov JP (2011) Integrative genomics viewer. Nat Biotechnol 29:24–26
23. Clinical and Laboratory Standards Institute (2024) Performance Standards for Antimicrobial Susceptibility Testing, 34th ed. (M100 ED34). Clinical and Laboratory Standards Institute, Wayne, PA
24. Kohanski MA, DePristo MA, Collins JJ (2010) Sublethal antibiotic treatment leads to multidrug resistance via radical-induced mutagenesis. Mol cell 37:311–320
25. Oz T, Guvenek A, Yildiz S, Karaboga E, Tamer YT, Mumcuyan N, Ozan VB, Senturk GH, Cokol M, Yeh P, Toprak E (2014) Strength of selection pressure is an important parameter contributing to the complexity of antibiotic resistance evolution. Mol Biol Evol 31:2387–2401
26. O'Leary NA, Wright MW, Brister JR, Ciufo S, Haddad D, McVeigh R, Rajput B, Robbertse B, Smith-White B, Ako-Adjei D, Astashyn A,

Badretdin A, Bao Y, Blinkova O, Brover V, Chetvernin V, Choi J, Cox E, Ermolaeva O, Farrell CM, Goldfarb T, Gupta T, Haft D, Hatcher E, Hlavina W, Joardar VS, Kodali VK, Li W, Maglott D, Masterson P, McGarvey KM, Murphy MR, O'Neill K, Pujar S, Rangwala SH, Rausch D, Riddick LD, Schoch C, Shkeda A, Storz SS, Sun H, Thibaud-Nissen F, Tolstoy I, Tully RE, Vatsan AR, Wallin C, Webb D, Wu W, Landrum MJ, Kimchi A, Tatusova T, DiCuccio M, Kitts P, Murphy TD, Pruitt KD (2016) Reference sequence (RefSeq) database at NCBI: current status, taxonomic expansion, and functional annotation. Nucleic Acids Res 44(D1):D733–D745

27. Martin M (2011) Cutadapt removes adapter sequences from high-throughput sequencing reads. EMBnet.journal 17(1):10–12
28. Andrews S (2010) FastQC: a quality control tool for high throughput sequence data. http://www.bioinformatics.babraham.ac.uk/projects/fastqc
29. Turner R (2024) deldir: Delaunay triangulation and Dirichlet (Voronoi) tessellation (v. 2.0-4). CRAN
30. Berdejo D, Merino N, Pagán E, García-Gonzalo D, Pagán R (2020) Genetic variants and phenotypic characteristics of *Salmonella* typhimurium-Resistant mutants after exposure to carvacrol. Microorganisms 2020:8

Chapter 4

Genetic Characterization of Foodborne Pathogens by Pulsed Field Gel Electrophoresis (PFGE): *Listeria Monocytogenes*

Belén Romero de Castilla, María Pilar Conchello and María Carmen Rota

Abstract

Molecular typing of pathogenic bacteria plays a crucial role in the food industry, enabling the detection, tracing, and control of foodborne pathogens such as *Listeria monocytogenes*. Among them, Pulsed Field Gel Electrophoresis (PFGE) plays an important role in the epidemiological characterization of these pathogens, identifying clonal relationships between isolates from food, environment, and clinical sources. This method has proven useful in outbreak investigation and understanding the pathogen persistence in food processing environments, contributing to the design of specific control strategies. This chapter details a step-by-step PFGE protocol for *Listeria monocytogenes* based on the standardized PulseNet protocol, providing detailed explanations of critical parameters to ensure accuracy and reproducibility.

Keywords PFGE, Molecular typing, Bacterial typing techniques, Genetic characterization, Foodborne pathogens, *Listeria monocytogenes*

1 Introduction

The presence and persistence of pathogenic bacteria in the food industry represent critical challenges to food safety and public health [1]. These microorganisms can contaminate food at different production stages, from raw materials to final products, posing a significant risk of foodborne illnesses. Pathogen persistence in food processing environments is particularly concerning due to their ability to form biofilms resistant to standard cleaning and disinfection procedures, becoming recurrent sources of contamination [1, 2]. This persistence, coupled with their ability to adapt to adverse conditions, represents a continuous risk to food safety.

One particularly concerning food pathogen for food safety is *L. monocytogenes*. Responsible for listeriosis, a serious disease in at-risk groups, *L. monocytogenes* accounted for 28 of the 64 food outbreak-related deaths reported in 2022 in the European Union, underscoring its particular virulence [3]. This pathogen exhibits

Marciane Magnani and Geany Targino de Souza Pedrosa (eds.), *Detection and Quantification of Bacteria, Yeast, Viruses, and Protozoan in Foods and Freshwater*, Methods and Protocols in Food Science, https://doi.org/10.1007/978-1-0716-5064-6_4,

adaptive mechanisms enabling survival in adverse conditions, such as low temperatures, wide pH ranges (4.6–9.5), high salinity (up to 20%), and limited water availability ($a_w < 0.90$). Its remarkable adaptability and biofilm-forming ability make it difficult to control this pathogen in processing plants and facilitate its persistence [4].

In this context, molecular typing of pathogenic bacteria has acquired significant importance in the food industry and public health. This tool is essential to investigate and control foodborne disease outbreaks, monitor the pathogen presence and persistence in production environments, identify microbial niches or reservoirs, detect and trace specific strains along the food chain, and establish epidemiological connections between isolates from different sources [5, 6].

Among the different molecular typing techniques, Pulsed-Field Gel Electrophoresis (PFGE) stands out for its high discriminatory power, versatility, and availability of standardized protocols that allow global inter-laboratory comparability [7]. Despite the emergence of advanced techniques such as whole genome sequencing, PFGE remains a reference technique in many contexts due to its robustness, cost-effectiveness, and the extensive database of genetic profiles accumulated over the years. The technique's adoption by international networks like PulseNet, coordinated by the United States Centers for Disease Control and Prevention (CDC), exemplifies its importance [8]. PulseNet has developed standardized PFGE protocols for pathogens of significant health concern, including *Campylobacter jejuni*, *Clostridium botulinum*, *Escherichia coli* (O157:H7 and non O157), *Salmonella* spp., *Shigella sonnei* and *Shigella flexneri*, *Listeria monocytogenes*, *Vibrio cholerae* and *Vibrio parahemolyticus*, and *Yersinia pestis* [9–14].

PFGE, developed by Schwartz and Cantor in 1984, is a molecular typing technique that allows efficient separation of large DNA fragments (up to 6 Mb in length) [15]. This technique is based on the reorientation of DNA molecules in response to periodic changes in the direction of the electric field applied to an agarose gel. The reorientation speed depends on the size of the molecules. Smaller molecules reorient faster; therefore, the migration of larger molecules through the gel is slower and, consequently, the molecules separate from each other [15, 16].

This technique is based on immobilizing intact chromosomal DNA in agarose plugs. For this purpose, cells are cultured, and the cell concentration is adjusted, mixed with agarose, and transferred into a suitable mold to form the plugs. Then, cells are lysed, and DNA is deproteinized in situ by incubating the agarose plugs with specific chemical reagents. Subsequently, restriction endonucleases that recognize 6–8 base pair sequences are used to generate DNA fragments that will be separated during electrophoresis [7, 17, 18].

DNA fragment separation in PFGE highly depends on the electrode system used [7]. The Contour-clamped Homogeneous Electric Field (CHEF) system is the most used configuration. CHEF employs hexagonally positioned electrodes to generate two alternating homogeneous electric fields across the gel at a constant reorientation angle of 120° (Fig. 1). This configuration ensures a consistent DNA migration, achieving straight migration and optimal resolution, which facilitates comparison between different fingerprint patterns [19].

PFGE enables the discrimination of specific fingerprint patterns, allowing for the construction of unique genetic profiles. The applications of this technique are diverse, including determination of chromosomal DNA size in yeast, fungi, and unicellular parasites, and the study of genetic organization [20–22]. Nevertheless, the most widespread application is currently the typing of bacterial DNA.

The resulting fingerprint patterns determine bacterial clonal assignment. Fingerprint patterns are compared with those of other characterized isolates using standardized software, and a dendrogram is produced to assess the relationship between bacteria visually. Isolates whose fingerprint patterns are indistinguishable (>95%) are considered the same strain, while those that are "closely related" (>70%) are considered part of the same cluster [23]. This technique enables the comparison and tracing of strains during outbreak investigations and persistence studies in food processing environments.

This versatility makes PFGE an indispensable tool for outbreak investigations and studies on pathogen persistence in food processing environments. While reagents, lysis conditions, and electrophoresis parameters depend on the bacterial genus, standardized protocols ensure reproducibility and comparability across studies [9–14]. This chapter focuses on the PFGE protocol for *L. monocytogenes*, based on the protocol standardized by PulseNet [12].

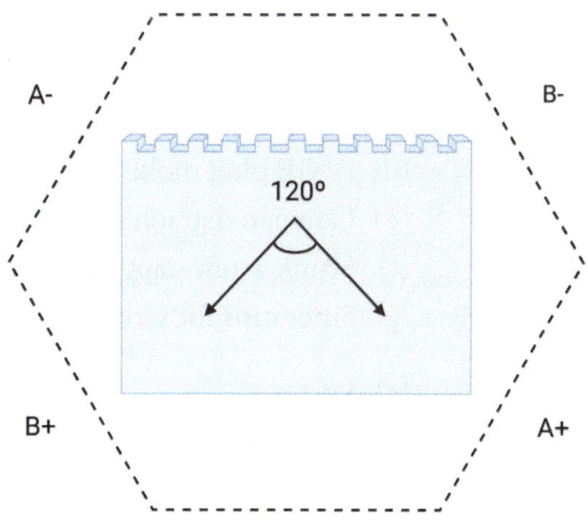

Fig. 1 Schematic illustration of the CHEF configuration used in PFGE. The outer hexagon represents the electrophoresis cuvette. Symbols A+ and A-, B+ and B- indicate the positions of the electrode pairs of the alternating electric fields. The central rectangle represents the agarose gel with its corresponding wells. The arrows show the electric field vectors of the alternating electric fields at an angle of 120 degrees

2 Materials

2.1 Growth Media

Prepared according to manufacturer instructions:

(a) Brain Heart Infusion Broth (BHI).
(b) Brain Heart Infusion Agar (BHIA).

2.2 Bacterial Strains

(a) Cryovial of the bacterial strain to study.
(b) *Salmonella* Braenderup H98122.

2.3 Reagents

(a) Buffers: TE Buffer (Tris 10 mM, EDTA 1 mM, pH 8.0); Cell Suspension Buffer (CSB, 100 mM Tris, 100 mM EDTA, pH 8.0); Cell Lysis Buffer (50 mM Tris, 50 mM EDTA, pH 8.0 + 1% Sarcosyl); TBE Buffer (Tris-Borate-EDTA 10×, pH 8.0).
(b) Sodium Dodecyl Sulfate (SDS, ≥99.0%).
(c) N-lauroylsarcosine sodium salt (Sarcosyl, 95%).
(d) Seakem® Gold Agarose (SGA).
(e) Lysozyme (20 mg/mL 1× TE buffer).
(f) Proteinase K (20 mg/mL CLRW).
(g) Enzymes *Asc*I, *Apa*I, and *Xba*I and restriction buffer.
(h) Bovine Serum Albumin (BSA).
(i) GelRed® Nucleic Acid Stain.

2.4 Laboratory Supplies

(a) Tools: inoculation loops, micropipettes, pipettes, pipette tips, plastic 1.5-mL tubes, scalpel, glass Pasteur pipette, Petri dishes (90 mm), large microscope slides.
(b) 16 mm diameter glass tubes.
(c) 100 mL screw-cap glass flask.
(d) PFGE plug molds.
(e) Laminated graph paper.
(f) 50 mL screw-capped polypropylene tubes.
(g) Filter caps for screw-capped polypropylene tubes.

2.5 Equipment

(a) Vortex.
(b) Densitometer.
(c) Water bath with stirring.
(d) Thermoblock.
(e) CHEF-DR III instrument.
(f) Orbital shaker.
(g) Covered metal container.
(h) Gel Doc XR+ Gel Documentation System.

2.6 Software

(a) GelDoc XR+.

(b) Quantity One Analysis Software.

(c) GelCompar II.

3 Methods

3.1 Preparation of Bacterial Strains and Cell Suspensions

1. Inoculate a 10 mL BHI broth test tube with an inoculation loop from the cryovial of the bacterial strain to study (*see* **Note 1**).
2. Homogenize in a vortex and incubate the test tubes at 37°C for 24 h.
3. Inoculate and streak over the BHIA plate's surface with an inoculation loop to obtain single colonies (*see* **Note 1**).
4. Incubate the agar plates at 37°C for 24 h.
5. Inoculate and streak over a single colony on the BHIA plate's surface with an inoculation loop to obtain single colonies and ensure a pure culture.
6. Incubate the agar plates at 37°C overnight (14–18 h).

   Repeat this process with all the bacterial strains to be studied and the reference strain *Salmonella* Braenderup II98122.
7. Label 16 mm diameter glass tubes with the identification code of each isolate (*see* **Note 2**).
8. Transfer 2 mL of 1× TE buffer to labeled tubes for *Listeria* isolates and 2 mL of Cell Suspension Buffer for *Salmonella* Braenderup H98122 standard.
9. Resuspend isolated colonies from BHIA plates in the corresponding labeled test tubes with an inoculation loop and gently swirl until completely dissolved.
10. Measure the concentration of the cell suspensions using a densitometer and adjust to 5.8–6.3 McFarland (16-mm-diameter tubes) (*see* **Note 3**).
11. Transfer 400 μL of the bacterial suspension to a plastic 1.5-mL tube. Test tubes can be kept refrigerated if the process needs to be repeated.

3.2 Preparation of PFGE Plugs

A. Agarose preparation

   1. Heat a plastic tube with 1.5 mL of a 10% solution of Sodium Dodecyl Sulfate (SDS) in a water bath previously heated to 55°C.
   2. Weigh 0.25 g Seakem Gold Agarose (SGA) into a 100 mL screw-cap glass flask, add 25 mL of 1× TE buffer, and swirl gently.

3. Dissolve the agarose, with the flask cap loosened, by heating in the microwave at short intervals (10–30 s), stirring manually until completely dissolved.
4. Add 1.25 mL of the heated 10% SDS solution and homogenize gently using rotary motions (to avoid forming bubbles).
5. Keep in a water bath at 55°C until use (minimum 15 min).

B. Preparation of PFGE plugs

1. Label the PFGE plug molds with each isolate code (four molds per isolate). Reusable or single-use plug molds can be used.

For *Listeria* spp. isolates:

2. Add 20 μL of thawed lysozyme (20 mg/mL) to each plastic 1.5-mL tube with 400 μL of the bacterial suspension, and mix gently with a pipette. Discard unused thawed lysozyme solution.
3. Incubate in a water bath at 55°C for 20 min.

For *Listeria* spp. and *Salmonella* Braenderup H98122 isolates:

4. Add 20 μL of proteinase K (20 mg/mL) to each plastic 1.5-mL tube of the bacterial suspension and mix gently with a pipette.
5. Transfer 400 μL of the bacterial suspension with lysozyme and proteinase K to new labeled plastic 1.5-mL tubes.
6. Add 400 μL of pre-heated agarose (1% SGA + 0.5% SDS) to 55°C, pipetting gently up and down 2–3× maximum. Over-pipetting may cause denaturation of the DNA (*see* **Note 4**).
7. Place the flask with the agarose back into the water bath to maintain its temperature.
8. Immediately after mixing the agarose with the bacterial suspension, dispense an appropriate amount of the mixture into each PFGE plug mold. Place the pipette tip on the bottom of the plug mold and pipetting gently to avoid bubble formation. The volume will depend on the plug mold used. With single-use molds, up to four plugs can be formed per sample, and with reusable molds, up to two plugs (*see* **Note 5**).
9. Allow the PFGE plugs to solidify at room temperature for 15 min.

3.3 Lysis of Cells in Agarose Plugs

1. Label 50 mL polypropylene screw-cap tubes with the code of each isolate.

2. Prepare the total volume of Cell Lysis Buffer/proteinase K (5 mL:25 μL) required for all tubes ($n+1$) in an appropriately sized test tube or flask and aliquot 5 mL to each 50 mL labeled polypropylene screw-cap.
3. Trim excess agarose from the top of the PFGE plugs with a scalpel or similar instrument. This step is optional, but it is recommended to obtain a homogeneous size of the plugs.
4. Open the reusable PFGE plug mold and transfer the plugs from the mold with a spatula to the labeled polypropylene screw-cap tube. If using disposable plug molds, remove the white adhesive tape from the bottom and push the plugs with the plastic mold spatula into the labeled polypropylene screw-cap tube. Be sure that the plugs are dipped into the buffer.
5. Place the polypropylene screw-cap tubes in a metal rack and incubate in a shaking water bath at 54–55°C for 2 h with constant vigorous shaking at 175–200 revolutions per min (rpm). Be sure that the water level of the water bath is above the level of the lysis buffer in the tubes.
6. Place the PFGE plug mold, spatulas, and scalpel in a suitable disinfectant. Soak for at least 30 min before washing. Properly dispose of the disposable molds.

3.4 Washing of Agarose Plugs After Cell Lysis

1. Pre-heat enough sterile ultrapure water and 1× TE buffer at 54–55°C to wash the plugs after lysis (30 mL sterile ultrapure water/tube and 60 mL 1× TE buffer/tube).
2. Remove polypropylene screw-cap tubes from the water bath and fit filter caps to tubes (*see* **Note 6**). Pour off lysis buffer by decanting into a container with a suitable disinfectant and dry the filter cap with laboratory tissues (to avoid leaving lint residues). If no filter cap is used, assist with a spatula.
3. Add 15 mL of pre-heated ultrapure water at 54°C to each polypropylene screw-cap tube and incubate at 54°C for 10 min in a water bath with constant vigorous shaking at 175–200 rpm.
4. Pour off water and dry the filter cap with laboratory tissues.
5. Repeat Steps 3 and 4 one more time.
6. Add 15 mL of pre-heated 1× TE buffer at 54°C to each polypropylene screw-cap tube and incubate at 54°C for 15 min in a water bath with constant vigorous shaking at 175–200 rpm.
7. Pour off 1× TE buffer and dry the filter cap with laboratory tissues.
8. Repeat Steps 6 and 7 three times.
9. Label new 50 mL polypropylene screw-cap tubes with the isolate codes, add 10 mL of 1× TE buffer to each, and place the

plugs with a blunt-tipped glass Pasteur pipette or spatula into the new polypropylene screw-cap tubes.

10. Store polypropylene screw-cap tubes refrigerated until use. TE-buffered plugs can be stored for up to 1 year under refrigeration; therefore, it is recommended to transfer to smaller tubes for long-term storage.

3.5 Restriction Digestion of DNA in Agarose Plugs

*Asc*I is used as the primary enzyme for *Listeria* spp. restriction because less amount of the enzyme is required. *Apa*I is recommended as a secondary enzyme in situations where PFGE patterns obtained with the primary enzyme are indistinguishable.

The enzyme used for the restriction digestion of *Salmonella* Braenderup H98122 is *Xba*I.

1. Label plastic 1.5-mL tubes with isolate codes and three or four tubes for *Salmonella* Braenderup H98122 (for 10- or 15-well gel, respectively). *Salmonella* Braenderup H98122 standards should be placed in the end wells of the gel and between samples. No more than four samples should be placed between the standard lanes.
2. A pre-restriction step (incubation step before restriction) is highly recommended. Prepare enough pre-restriction buffer for all plugs ($n+1$) in plastic 5-mL tubes by diluting the appropriate 10× restriction buffer with sterile ultrapure water (1:10) (Table 1) (*see* **Note 7**).
3. Add 200 μL diluted pre-restriction buffer (1×) to labeled plastic 1.5-mL tubes.
4. Carefully remove plugs from 1× TE buffer with a spatula or blunt-tipped glass Pasteur pipette and place in a sterile disposable Petri dish or large glass slide.
5. Cut the plug in half lengthwise if using the 10-well gel or in three portions vertically if using the 15-well gel, and the appropriate number of *Salmonella* Braenderup H98122 with a scalpel or similar utensil (*see* **Notes 8** and **9**).
6. Transfer the plug slice into a plastic 1.5-mL tube containing the pre-restriction buffer. Be sure that the plug slice is dipped into the buffer. Replace the rest of the plug back in original tube with TE buffer. Plugs should be stored at 4°C.

Table 1
Composition of pre-restriction buffer

Reagent	μL/plug
Sterile ultrapure water	180
10× restriction buffer	20

7. Incubate sample and *Salmonella* Braenderup H98122 plugs in a water bath or thermoblock for 10 min: samples to be digested with *Asc*I and *Xba*I at 37°C and with *Apa*I at 30°C (Roche/Sigma) or 25°C (New England BioLabs).
8. After incubation, remove the buffer by placing a 200–250 µL tipped pipette at the bottom of the tube and aspirate the buffer. Take care not to damage the plug slice with the pipette tip and not to absorb the plug slice with the pipette tip.
9. Prepare the restriction enzyme master mix (Tables 2, 3, and 4) (*see* **Note 10**).
10. Add 200 µL of restriction master mix to each plastic 1.5-mL tube. Ensure that the plugs are immersed in the enzyme mixture.

Table 2
Composition of *Asc*I restriction enzyme master mix

Reagent	µL/plug
Sterile ultrapure water	176.5
10× restriction buffer	20
BSA (20 mg/mL)	1
*Asc*I (10 U/µL)	2.5

Table 3
Composition of *Apa*I restriction enzyme master mix

Reagent	µL/plug
Sterile ultrapure water	178.5
10× restriction buffer	20
BSA (20 mg/mL)	1
*Apa*I (50 U/µL)	0.5

Table 4
Composition of *Xba*I restriction enzyme master mix

Reagent	µL/plug
Sterile ultrapure water	176.5
10× restriction buffer	20
BSA (20 mg/mL)	1
*Xba*I (20 U/µL)	2.5

11. Incubate plugs for 2 h in a water bath or thermoblock at the appropriate temperature for the enzyme: samples restricted with *Apa*I at 25°C and with *Asc*I and *Xba*I at 37°C.
12. After incubation, repeat Step 8.
13. Add 200 μL 0.5× TBE buffer to each plastic 1.5-mL tube with the plug and incubate at room temperature for 5 min. Digested plugs can be stored under refrigeration in 0.5× TBE for up to 3 days.
14. After incubation, repeat Step 8.

3.6 Prepare the PFGE Equipment CHEF-DR III

1. Check that the electrophoretic cuvette is leveled (midpoint and corners) in the CHEF-DR III equipment. Adjust if necessary (*see* **Note 11**). Check that all connections are correct (Fig. 2)
2. Pour 2.3 L ultrapure water into the electrophoresis cuvette for cleaning. Turn on the CHE-DR-III and the flow pump at 70 rpm and let it circulate for 20 min.
3. Turn off the flow pump and place a container on the floor with a capacity of more than 2 L at the height of the drainage tube outlet.
4. Remove the plastic cooling module tube from the electrophoresis cuvette. Place the drain tube in drain outlet and insert the tube outlet into the container on the floor.
5. Turn on the flow pump and remove the wash ultrapure water, tilting the trough if necessary.
6. Take off the drain tube from the electrophoresis cuvette and reconnect the cooling module tube.
7. Pour 2.3 L 0.5× TBE buffer into the electrophoresis cuvette.

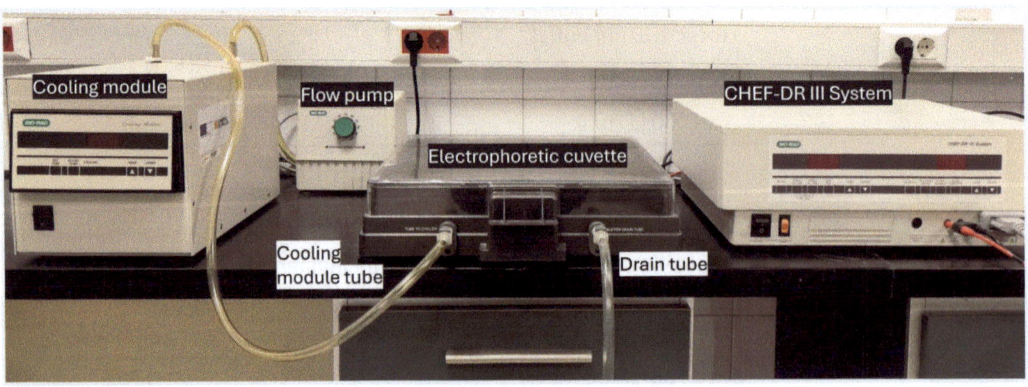

Fig. 2 CHEF-DR III equipment with labeled components. The setup includes a cooling module, flow pump, electrophoretic cuvette, and CHEF-DR III system. The cooling module is connected to the electrophoretic cuvette via cooling module tube and to the flow pump. The electrophoretic cuvette features a drain tube connector

8. Turn on the flow pump at 70 rpm and the cooling module at 14°C and allow to circulate for at least 30 min before introducing the gel.

3.7 Preparation of Agarose Gel

1. Turn on the water bath at 55°C.
2. Weigh 1 g SGA into a 250 mL screw-cap glass flask and add 100 mL 0.5× TBE buffer.
3. Dissolve the agarose, with the flask cap loosened, by heating in the microwave at short intervals (15 s), stirring manually until completely dissolved.
4. Transfer 1.5 mL of dissolved agarose to a plastic tube and maintain the plastic tube and screw-cap glass flask in the 55°C water bath until use with gentle agitation (minimum 15 min).
5. Assemble the gel form, verifying that the middle and corner points are leveled.
6. Place the comb flat on gel form, remove the plug slices from the plastic 1.5-mL tubes, and place plug slices on the bottom edge of the teeth. Include *Salmonella* Braenderup H98122 standards in outside lanes (teeth) and between samples as needed (no more than 4 samples between standards). Load samples on the remaining teeth of the comb and note locations (*see* **Note 12**).
7. Remove excess buffer with laboratory tissues. Allow the plug slides to air dry on the comb for 3–5 min.
8. Place the comb in leveled gel mold and confirm that the plug slices are correctly aligned on the bottom of the comb teeth and that the bottom edge of the plug slice is flush against the black platform.
9. Carefully pour the melted agarose (tempered to 55°C) from the screw-cap glass flask into the gel form.
10. Allow to solidify at room temperature for 30 min (*see* **Note 13**).
11. Once the gel has solidified, carefully remove the comb and fill the gel wells with the melted agarose (tempered to 55°C) from the plastic 1.5-mL tube.
12. Allow to solidify at room temperature for 5 min (*see* **Note 13**).
13. Demold the gel and clean the agarose residues from the edges and the bottom of the platform with laboratory tissues.

3.8 Electrophoresis

1. Place the gel in the black frame inside the cuvette, close the lid, and allow it to equilibrate for 10 min before starting electrophoresis.
2. Enter the conditions into the CHE-DR III instrument (Table 5) and start electrophoresis (*see* **Note 14**).

Table 5
Electrophoresis conditions in the CHEF-DR III instrument

Parameter	Value
Initial switching time	4.0 s
Final switching time	40.0 s
Voltage	6 V
Included angle	120°
Operating time	18.8 h

3. Record the initial milliamp (mA) reading on the instrument. The initial mA should be between 110 and 150 mA. A reading outside this range may indicate that the 0.5× TBE buffer was prepared incorrectly and should be remade.

3.9 Gel Staining and Image Acquisition

1. After electrophoresis, turn off the cooling module, flow pump, and equipment. In a metal covered container prepare the staining solution (*see* **Note 15**): 250 mL of distilled water and 75 μL of GelRed®.
2. Place the covered metal container on an orbital shaker.
3. Remove the gel from the electrophoretic cuvette and place it in the metal container. Do not insert the black base of the mold into the staining solution.
4. Check that all the gel is immersed and switch on the orbital shaker at 30 rpm.
5. Stain the gel for 30 min.
6. After staining, switch off the orbital shaker and, using a spatula, remove the gel from the staining solution and the excess liquid using laboratory tissues.
7. Place the gel on the platform of the image acquisition equipment to be used (GelDoc XR, XR+, or equivalent documentation system).
8. Open the image acquisition software on the computer linked to the instrument, check that the gel is framed, and use the software zoom to fit the gel image to the window. Adjust if necessary. The gel image should fill the imaging screen window (without cutting off wells or lower bands).
9. Once the position and focus are correct, turn on the UV light in the image acquisition equipment. Make sure the image is in focus and that there is little or no saturation (overexposure) in the bands (indicated by red pixels in the software).
10. Capture images according to the instructions provided with your image acquisition software. Save the gel image as a *.1sc

(QuantityOne software), *.scn (ImageLab software), or *.tif (BioNumerics software) file.

11. Remove the gel from the image acquisition equipment using laboratory tissues or a spatula and dispose of it promptly.
12. Clean the platform and the exterior of the image acquisition equipment with the appropriate disinfectant.

3.10 Gel Image Analysis

1. Open GelCompar II software.
2. Create a new database (local or connected) (*see* **Note 16**), configure the information fields, and manually add entries or import files.
3. Configure a fingerprint experiment type by selecting parameters for gel image processing, such as TIFF file format and densitometric values (*see* **Note 17**).
4. Import gel images and convert them to grayscale.
5. Outline gel lines area, define lanes, adjust image settings for band enhancement, and enhance image quality by applying Least Square Filtering and Background Subtraction.
6. Assign the molecular weight to each band of a *Salmonella* Braenderup H98122 reference line and normalize the image accordingly.
7. Search for bands in non-standard lines using a minimum profiling of 5 and a gray zone of 2.
8. Link fingerprint data to database entries and the standard lanes to reference entries.
9. Establish a calibration curve by assigning units (kb) and markers to the reference system.
10. Select non-standard entries and compare isolates using a similarity matrix using Dice's band-based similarity coefficient with 1% optimization and tolerance.
11. Generate a dendrogram using the Unweighted Pair Group Method with Arithmetic mean (UPGMA) to evaluate genetic relationships. The dendrogram shows the percentage of similarity between isolates analyzed based on bands. Isolates whose fingerprint patterns are indistinguishable (>95%) are considered the same strain, while those that are "closely related" (>70%) are considered part of the same cluster [23].

3.11 Cleaning the PFGE Equipment CHEF-DR III

1. Repeat Steps 3–6 from the Sect. 3.6. Prepare the PFGE Equipment CHEF-DR III.
2. Pour 2.3 L of ultrapure water into the electrophoresis cuvette for cleaning.
3. Turn on the CHE-DR III and flow pump at 70 rpm and let it circulate for 20 min.

4. Repeat Steps 3–6 from the Sect. 3.6. Prepare the PFGE Equipment CHEF-DR III.

4 Notes

1. Any non-selective nutrient medium can be used, but nutrient-rich media are recommended for optimal *Listeria* growth.
2. Bacterial concentration can be adjusted by spectrophotometer, turbidity meter, or colorimeter. The tubes used and the minimum volume of cell suspension required will depend on the size of the cuvettes or tubes and the specifications of the equipment used to measure the cell concentration.
3. The turbidity will depend on the diameter of the tube used and the microorganism to study. It is recommended to make a comparison the first time the protocol is performed, measuring the turbidity in a densitometer and spectrophotometer. These values gave satisfactory results in the Department of Animal Production and Food Technology of the University of Zaragoza; each laboratory may need to establish the optimum concentration necessary to obtain satisfactory results.
4. Plug formation should be done as quickly as possible to minimize the solidification of the agarose in the pipette tips/plastic 1.5-mL tubes. Addition of agarose and plug formation should be done one at a time.
5. Keep the plastic 1.5-mL tubes with the suspension and agarose in hand while dispensing into the molds to maintain temperature.
6. The use of filter caps is optional but recommended to speed up the washing process.
7. The appropriate restriction buffer varies among vendors. Always use the vendor's recommended restriction buffer for that restriction enzyme.
8. It is recommended to have laminated graph paper to facilitate the plug cutting.
9. The computer analysis of the gels is more accurate and less tedious when using the 10-well comb; therefore, it is recommended to use the 10-well comb until experienced in the technique.
10. Enzymes can be obtained at different stock concentrations. Calculations are based on the concentrations shown in Tables 2, 3, and 4. In case of using a different concentration, make the necessary adjustments to reach a final concentration of 25 U/sample, with the volume of enzyme and water. Also, the

Bovine Serum Albumin (BSA) concentration is calculated for an initial concentration of 20 mg/mL. If a different concentration is used, make BSA and water volume adjustments to achieve a final concentration of 0.1 mg/mL.

11. Electrophoresis can be carried out using various equipment (CHEF Mapper, CHEF-DR II, and CHEF-DR III). The volumes used as well as the electrophoresis conditions will depend on the equipment used.
12. Alternatively, plug cuts can be loaded into the gel by first forming the agarose gel and then inserting the plug cuts into the wells. If this procedure is followed, it is important to very carefully push the plug to the bottom of the well. Each laboratory can develop its technique for systematically loading the plug slides into the wells so that the lanes are straight, and the bands are neat.
13. It is recommended to cover the gel with aluminum foil or similar to prevent particles from falling into the gel during solidification.
14. Electrophoresis time is based on the equipment and reagents used. The ambient temperature strongly influences this time. Run times may be different in your laboratory and will need to be optimized for your gels so that the lowest band in the standard migration of *Salmonella* Braenderup H98122 is between 1.0 and 1.5 cm from the bottom of the gel.
15. Other alternative DNA stains may be used. The volumes described may vary depending on the size of the container used.
16. A local database is recommended when only one user will use the database. The connected database is recommended in environments where many users need to access the database simultaneously or access to an existing database is available.
17. The densitometric values applied depend on the specific stain used.

References

1. Møretrø T, Langsrud S (2017) Residential bacteria on surfaces in the food industry and their implications for food safety and quality. Compr Rev Food Sci Food Saf 16:1022–1041
2. Srey S, Jahid IK, Ha S-D (2013) Biofilm formation in food industries: a food safety concern. Food Control 31:572–585
3. European Food Safety Authority (EFSA), European Centre for Disease Prevention and Control (ECDC) (2023) The European Union One Health 2022 Zoonoses report. EFSA J 21:e8442
4. Osek J, Lachtara B, Wieczorek K (2022) *Listeria monocytogenes* - how this pathogen survives in food-production environments? Front Microbiol 13:866462
5. Alegbeleye OO, Sant'Ana AS (2020) Pathogen subtyping tools for risk assessment and management of produce-borne outbreaks. Curr Opin Food Sci 32:83–89

6. Kovac J, Bakker H den, Carroll LM, Wiedmann M (2017) Precision food safety: a systems approach to food safety facilitated by genomics tools. TrAC Trends Anal Chem 96:52–61
7. Lopez-Canovas L, Martinez Benitez MB, Herrera Isidron JA, Flores Soto E (2019) Pulsed field gel electrophoresis: past, present, and future. Anal Biochem 573:17–29
8. Ribot EM, Hise KB (2016) Future challenges for tracking foodborne diseases: PulseNet, a 20-year-old US surveillance system for foodborne diseases, is expanding both globally and technologically. EMBO Rep 17:1499
9. PulseNet (2013) Standard operating procedure for PulseNet PFGE of *Campylobacter jejuni* (PNL03). https://www.pulsenetinternational.org/assets/PulseNet/uploads/pfge/PNL03_CampyPFGEprotocol.pdf. Accessed 15 Nov 2024
10. PulseNet (2010) Standard operating procedure for PulseNet PFGE of *Clostridium botulinum* (PNL25). https://www.pulsenetinternational.org/assets/PulseNet/uploads/pfge/CBotulinum2010.pdf. Accessed 15 Nov 2024
11. PulseNet (2013) Standard operating procedure for PulseNet PFGE of *Escherichia coli* O157:H7, *Escherichia coli* non-O157 (STEC), *Salmonella* serotypes, *Shigella sonnei* and, *Shigella flexneri* (PNL05). https://www.pulsenetinternational.org/assets/PulseNet/uploads/pfge/PNL05_Ec-Sal-ShigPFGEprotocol.pdf. Accessed 15 Nov 2024
12. PulseNet (2013) Standard operating procedure for PulseNet PFGE of *Listeria monocytogenes* (PNL04). https://www.pulsenetinternational.org/assets/PulseNet/uploads/pfge/PNL04_ListeriaPFGEprotocol.pdf. Accessed 15 Nov 2024
13. PulseNet (2013) Standard operating procedure for PulseNet PFGE of *Vibrio cholerae* and *Vibrio parahaemolyticus* (PNL06). https://www.pulsenetinternational.org/assets/Uploads/PNL06-Vchol-Vparah-PFGE-protocol.pdf. Accessed 15 Nov 2024
14. PulseNet (2006) One-Day (24–28 h) Standardized laboratory protocol for molecular subtyping of *Yersinia pestis* by pulsed field gel electrophoresis (PFGE). https://www.pulsenetinternational.org/assets/PulseNet/uploads/pfge/YpestisPFGEprotocol.pdf. Accessed 15 Nov 2024
15. Schwartz DC, Cantor CR (1984) Separation of yeast chromosome-sized DNAs by pulsed field gradient gel electrophoresis. Cell 37:67–75
16. Stellwagen NC (2009) Electrophoresis of DNA in agarose gels, polyacrylamide gels and in free solution. Electrophoresis 30:S188–S195
17. Smith CL, Cantor CR (1987) Preparation and manipulation of large DNA molecules: advances and applications. Trends Biochem Sci 12:284–287
18. Birren B, Lai E (1993) Chapter 3. Preparation of DNA for pulsed field analysis. In: Birren B, Lai E (eds) Pulsed field gel electrophoresis: a practical guide. Academic, San Diego, 25–74
19. Chu G, Vollrath D, Davis RW (1986) Separation of large DNA molecules by contour-clamped homogeneous electric fields. Science 234:1582–1585
20. Hage AE, Houseley J (2013) Resolution of budding yeast chromosomes using Pulsed-Field Gel Electrophoresis. In: Makovets S (ed) DNA electrophoresis: methods and protocols. pp 195–207
21. Souza RT, Lima FM, Barros RM, Cortez DR, Santos MF, Cordero EM, Ruiz JC, Goldenberg S, Teixeira MMG, Silveira JFda (2011) Genome size, karyotype polymorphism and chromosomal evolution in *Trypanosoma cruzi*. PLoS One 6:e23042
22. Shizuya H, Kouros-Mehr H (2001) The development and applications of the bacterial artificial chromosome cloning system. Keio J Med 50:26–30
23. Martin P, Jacquet C, Goulet V, Vaillant V, De Valk H (2006) Participants in the PulseNet Europe Feasibility Study pulsed-field gel electrophoresis of *Listeria monocytogenes strains*: the PulseNet Europe Feasibility Study. Foodborne Pathog Dis 3:303–308

Chapter 5

Detection and Quantification of Marker and Non-Marker Genes of Interest in Cheese Microbiota by qPCR. Focus on Biogenic Amine Gene Profiles

Livia Balaguer, Natalia Merino, Diego García-Gonzalo, Rafael Pagán, Daniel Berdejo and Laura Espina

Abstract

Biogenic amines (BAs) are crucial indicators of microbial activity and pose potential safety risks in fermented foods such as cheese. This study focuses on developing and implementing a quantitative PCR (qPCR) method to detect and quantify *hdcA*, a non-marker gene associated with histamine bacterial biosynthesis, within the microbiota of cheese samples. The methodology includes conducting standard curves for both non-marker and marker genes before the qPCR experiments. These experiments amplify both types of genes in DNA samples from cheese microbiota, allowing for the interpolation of their abundances from the established standard curves. Abundances of the non-marker gene are then normalized against the conserved marker gene, the 16s rRNA gene, present in these samples. This approach offers a quick, easy, and reliable tool for detecting and quantifying *hdcA* gene in cheese. Therefore, it can be used as a convenient screening method to better understand and manage the risks linked to biogenic amines in fermented foods.

Keywords *hdcA*, qPCR, Cheese, Microbiome, Biogenic amines, Marker genes, Non-marker genes

1 Introduction

In recent years, the study of biogenic amines (BAs) in fermented foods has grown increasingly important, driven by their significant impact on both food quality and safety [1]. BAs are organic nitrogen compounds, typically formed by the microbial decarboxylation of amino acids during fermentation or spoilage processes [2]. Common BAs such as histamine, tyramine, putrescine, and cadaverine are frequently detected in foods like cheese, fish, wine, and cured meats. While these compounds contribute to the characteristic flavor profile of fermented foods, they also pose potential health risks when consumed in high amounts, especially for individuals with metabolic sensitivities [1, 3]. Cheese, as a naturally fermented dairy product, is

Marciane Magnani and Geany Targino de Souza Pedrosa (eds.), *Detection and Quantification of Bacteria, Yeast, Viruses, and Protozoan in Foods and Freshwater*, Methods and Protocols in Food Science, https://doi.org/10.1007/978-1-0716-5064-6_5,

particularly susceptible to BA accumulation. Its distinctive microbial community, which can include spoiling microorganisms, might lead to significant biogenic amine production, especially under specific fermentation and ripening conditions [4].

Microbial genomes harbor specific genes encoding enzymes responsible for amino acids decarboxylation, a key step in BAs formation. By studying these genes, researchers can gain insights into which microbial populations are predisposed to produce BAs under different conditions [5]. Histamine is considered one of the most significant biogenic amines found in cheese [6]. The *hdcA* gene, which encodes histidine decarboxylases, is linked to bacteria, particularly lactic acid bacteria (LAB) such as *Lentilactobacillus buchneri* or *Lentilactobacillus parabuchneri*, among other, especially in varieties where bacterial activity drives the ripening process [4, 7]. In this context, the molecular technique quantitative polymerase chain reaction (qPCR) is a precise targeted tool for detecting and quantifying specific genes such as *hdcA* [8], giving direct insight into their abundance within microbial communities. Moreover, quantifying the *hdcA* gene can establish a direct link between its copy number and the amount of histamine measured using high-performance liquid chromatography (HPLC). Additionally, qPCR provides a fast and straightforward method for result normalization, which is crucial for maintaining consistent correlations across diverse samples and techniques [8, 9]. This is achieved by measuring the copy number of the 16S rRNA gene, a gene commonly used as an indicator of total bacterial concentration, and expressing the abundance of *hdcA* as a ratio relative to the 16S rRNA copy number. This ratio provides a standardized measure of the target gene's abundance, adjusted for variations in bacterial load [10].

Therefore, in this chapter, we describe the use of qPCR for detecting and quantifying the *hdcA* gene. Additionally, we detail the process of result normalization by utilizing qPCR to measure the 16S rRNA gene, which serves as a reference point for estimating the relative abundance of *hdcA* (or other AB-related genes) in the samples.

2 Materials

2.1 Standard Curve Performing

2.1.1 DNA Extraction

1. Pure culture.
2. Sterile toothpicks or inoculation needles.
3. Milli-Q sterile water.
4. Vortex mixer.

5. Laboratory tools: micropipettes, tips, and plastic 1.5-mL tubes.
6. DNA extraction kit: GenElute™ Bacterial Genomic DNA Kit.
7. Bead beater.
8. Microcentrifuge.
9. DNA quantification: Qubit™ 4 fluorometer, Qubit™ assay tubes, and Qubit 1× dsDNA HS Assay™, or equivalent.

2.1.2 PCR, DNA Quantification, and qPCR

1. Milli-Q sterile water.
2. Thermal cycler.
3. PCR reagents (DNA Polymerase, deoxynucleotides (dNTPs), and buffer solution) (see Table 1).
4. Forward and reverse primers.
5. 0.2-mL PCR tubes.
6. Purification PCR product: commercial spin-column kits, enzymatic treatment (ExoSAP-IT), or magnetic beads.
7. DNA quantification: Qubit™ 4 fluorometer, Qubit™ assay tubes, and Qubit 1× dsDNA HS Assay™, or equivalent.
8. qPCR reagents (DNA polymerase, deoxynucleotides (dNTPs), buffer solution, and SYBR Green™) (see Table 3).
9. qPCR plates and plate seal.

Table 1
Example list of components for each PCR reaction

Reagent	Volume (μL) for 50 μL reaction	Final concentration
5× Phusion Buffer	10	1×
dNTPs (10 mM)	1	200 μM
Forward primer (10 μM)	2.5	0.5 μM
Reverse primer (10 μM)	2.5	0.5 μM
Template DNA (previous preparation)	10	10–50 ng
Phusion DNA polymerase (2 U/μL)	0.5	0.02 U/μL
Nuclease-free water	Added up to a total volume of 50	

2.2 Sample Collection and Homogenization

1. Clean environment.
2. Sterile knife.
3. Sterile tongs.
4. Analytical balance.

5. Buffered peptone water (BPW).
6. Sterile filter bags.
7. Peristaltic homogenizer (stomacher).
8. Sterile bottles.

2.3 DNA Cheese Sample Extraction and Quantification

1. Milli-Q sterile water.
2. 15-mL sterile Falcon tubes.
3. Centrifuge capable of reaching 10,000 RCF.
4. Sterile swabs.
5. Buffer peptone water (BPW).
6. Laboratory tools: micropipettes, tips, and plastic 1.5-mL tubes.
7. Microcentrifuge.
8. Vortex mixer.
9. DNA extraction kit: DNeasy PowerFood™ Microbial Kit or alternative kit.
10. Bead beater.
11. DNA quantification: Qubit™ 4 fluorometer, Qubit™ assay tubes, and Qubit 1× dsDNA HS Assay™, or equivalent.

3 Methods

3.1 Experimental Design

This study aims to investigate the presence of genes linked to biogenic amines (BAs), particularly those involved in histamine production in cheese. The analysis utilizes the qPCR technique, with normalization performed using a 16S rRNA standard curve to ensure accurate comparisons of bacterial load across samples. The genetic findings can be further complemented with direct measurements of histamine concentrations performed with HPLC [2, 11]. In this chapter, we describe the methodology to detect and quantify the *hdcA* gene. Nevertheless, this methodology can also be adapted to quantify other genes (such as *tyrDC*, which is associated with the production of tyramine, another relevant BA).

Figure 1 shows the experimental steps.

3.2 Development of a Standard Curve for the Quantification of a Target Gene

This chapter provides an example of how to perform a qPCR standard curve, which is essential for the precise determination of the gene copy number. To create this curve, the DNA of the specific fragment amplified by conventional PCR must be purified, quantified and serially diluted to create a series of known concentrations. Each dilution is run in the qPCR assay, and the resulting Ct (Cycle threshold) values are plotted against the logarithm of their respective concentrations.

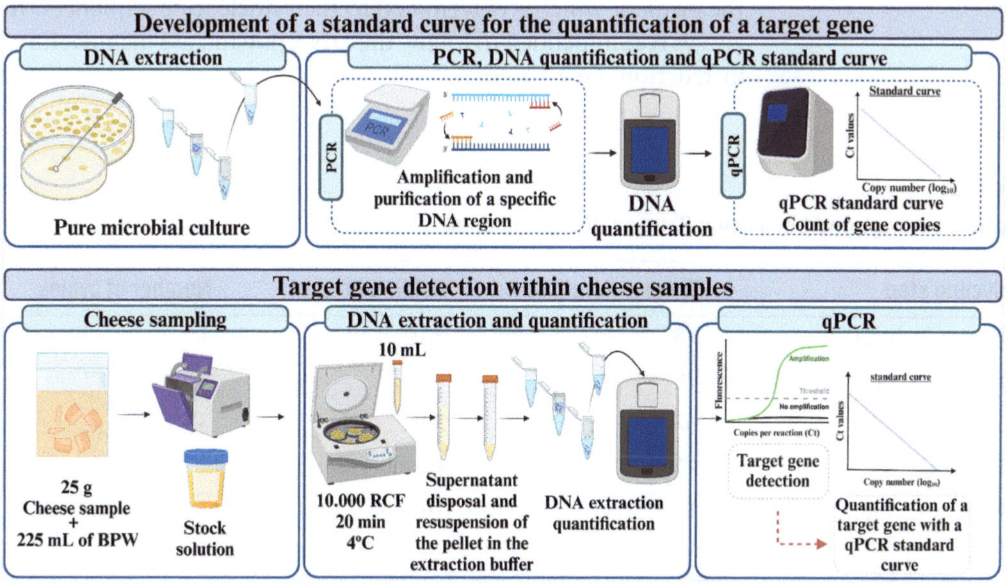

Fig. 1 Experimental steps to identify and quantify marker and non-marker genes by qPCR analysis. 1. Development of a standard curve for the quantification of a target gene. 2. Target gene detection within cheese samples

By interpolating the Ct values of unknown samples onto this curve, their initial gene concentrations can be accurately quantified.

3.2.1 DNA Extraction of Lentilactobacillus parabuchneri

DNA for pure culture can be obtained by selecting a bacterial strain known to contain the *hdcA* gene, such as *Lentilactobacillus parabuchneri* DSM 5987 [11]. It is recommended to perform a bacterial streak in de Man, Rogosa plates, and Sharpe agar medium (MRS agar) and incubate it during 72 h at 30°C under anaerobic conditions. Afterward, following the manufacturer's protocol for the GenElute™ Bacterial Genomic DNA Kit, select a few colonies and extract the DNA according to the provided instructions (*see* **Note 1**). Then, DNA is ready for the next experiments.

3.2.2 PCR Technique for Amplifying a Target Gene Region

The key components of the PCR are shown in Table 1.

The primer pair 5′-TTGACCGTATCTCAGTGAGTCCAT - 3′ and 5′- ACGGTCATACGAAACAATACCATC-3′ amplify the *hdcA* region [8], producing a 158 bp amplicon.

For each PCR reaction, the specific components, their volumes, and the order of addition depend on the specific DNA polymerase used.

The thermal cycler conditions of the PCR depend on the specific DNA polymerase and primers used in the reaction. In particular, the annealing temperature is chosen based on the melting temperature of the primers, which is determined by their nucleotide sequences. It is always recommended to follow the DNA polymerase manufacturer's instructions (see Table 2).

Table 2
Typical cycling conditions using Phusion DNA polymerase

Cycling step	Temperature (°C)	Holding time	Number of cycles
Enzyme activation	98	30 s	1
Denaturation	98	10 s	
Annealing	62	30 s	35
Extension	72	1.5 min (1 min/kb)	
Final extension	72	10 min	1
Hold	4	Until use	

3.2.3 Purification of the Amplicon

Purification of PCR products before qPCR analysis is crucial to remove unwanted components that might interfere with gene amplification. The aim is to eliminate primers, dNTPs, salts, and other contaminants. Common methods for PCR product purification include commercial spin-column kits, enzymatic treatment (ExoSAP-IT), or magnetic beads, any of which are suitable for this step. The DNA is now ready for subsequent experiments. At this stage, electrophoresis on an agarose gel is highly recommended to confirm whether the desired PCR product has been successfully amplified.

3.2.4 Quantification and Normalization of the DNA Concentration of the Amplicon

Measure the DNA concentration of the purified PCR fragment using the Qubit™ 4 fluorometer method or any other equivalent approach following the instructions (*see* **Note 2**). The DNA concentration should be adjusted to 25 ng in 20 μL (or another appropriate quantity depending on the specific experimental requirements) using Milli-Q sterile water.

3.2.5 Serial Dilutions of the Amplicon

Serial dilutions of the amplicon must be prepared using Milli-Q sterile water as the diluent, typically progressing in tenfold dilutions (*see* **Note 3**). Make at least 10 dilutions (for example, from 25 ng down to 25×10^{-10} ng). For each DNA concentration, the corresponding gene copy number can be calculated using the following formula:

$$Number\ of\ copies\ (molecules) = \frac{X\ ng \cdot 6.022 \cdot 10^{23} molecules/mol}{\left(N\left(bp\right) \cdot 660 \frac{g}{mol}\right) \cdot 1 \cdot 10^{9} ng/g}$$

where:

X = amount of amplicon (ng)

N = length of dsDNA amplicon (158 bp for *hdcA*)

660 g/mol = average mass of 1 bp dsDNA

6.022×10^{23} = Avogadro's constant

1×10^{9} = conversion factor from ng to g

Example calculation for the initial 25 ng point:

$$N^{o}\ of\ copies\ (molecules) = \frac{25ng \cdot 6.022 \cdot \frac{10^{23} molecules}{mol}}{\left(158 \cdot 660 \frac{g}{mol}\right) \cdot \frac{1 \cdot 10^{9} ng}{g}} = \mathbf{1.44 \cdot 10^{11} copies}$$

Each sequential dilution will have a copy number calculated in the same way as shown above. Once the operations are calculated, the experimental part is performed.

3.2.6 Conduction of qPCR for Each Amplicon Dilution

qPCR Plates

The key components of qPCR are detailed in Table 3 and the primers used are the same as described in Sect. 3.2.2.

It is recommended to use a SYBR Green™ Master Mix that includes DNA polymerase, dNTPs, buffer solution and SYBR Green™. The SYBR Green dye binds to double-stranded DNA and emits fluorescence when the DNA is amplified during the qPCR process. This fluorescence is monitored in real-time to quantify the DNA.

The steps to charge the plate are:

1. Prepare a master mix containing all the reagents as outlined before and add it into each well of the qPCR.
2. Add 1 μL of the sample from the corresponding dilution into each well.

Table 3
Example of list of components for each qPCR reaction

Reagent	Volume (μL) for 20 μL reaction	Final concentration
Master Mix SYBR Green™	10	1×
Forward primer (10 μM)	1	0.5 μM
Reverse primer (10 μM)	1	0.5 μM
Template DNA (previous preparation)	1	1.25 ng/μL
Nuclease-free water	7	

3. Seal the plate securely to prevent evaporation.
4. Ensure that all the content is at the bottom, there aren't air bubbles present in the well, and then proceed with running the qPCR.

qPCR Conditions

The thermal cycler conditions of the qPCR depend on the specific DNA polymerase and primers used in the reaction. In particular, the annealing temperature is chosen based on the melting temperature of the primers, which is determined by their nucleotide sequences. It is always recommended to follow the DNA polymerase manufacturer's instructions (see Table 4 for typical cycling conditions).

Table 4
Typical cycling conditions for qPCR amplification using *hdcA* primers

Cycling step	Temperature (°C)	Holding time	Number of cycles
Enzyme activation	95	3 min	1
Denaturation	95	15 s	35–40
Annealing/extension/plate read	60	30 s	
Melt curve	65–95	(increments of 0.5°C at 2–5 s/step)	
Hold	4	Until use	

3.2.7 Interpretation of the Ct Value

The Ct value is the number of cycles required for the fluorescent signal in a qPCR assay to cross a predefined threshold, indicating the presence of the target nucleic acid in the sample. The Ct value is inversely related to the amount of target gene present:

- Lower Ct values (<20): indicate a high concentration of the target gene, as the signal crosses the threshold quickly.
- Moderate Ct values (20–35): suggest a moderate concentration of the target.
- High Ct values (>35): often reflect a low target quantity, which may indicate very low nucleic acid levels or nonspecific amplification. Ct values near or above 40 are typically considered unreliable.

The threshold is a fluorescence level set above the baseline noise and below the plateau phase of amplification. Correct placement of the threshold is critical for accurate Ct determination. It is generally set within the exponential phase of amplification, where the signal is verifiably quantifiable over the noise.

For consistency and accurate normalization, it is recommended to use the same threshold across all samples within an experiment. Many qPCR systems provide automated algorithms to suggest a threshold, but it is essential to check and validate this setting to ensure reliable results. The Ct value is also referred to as Cq (Quantification cycle).

3.2.8 Plotting of the Standard Curve

Plot the $\log_{10}$ of the gene copy number (x-axis) calculated in Sect. 3.2.5 against the Ct values (y-axis) obtained doing the Sect. 3.2.7. Alternatively, the amount of DNA as starting material (in ng, as shown in Sect. 3.2.5) can be used in the x-axis. Fit a linear regression line to the data and check the R^2 value (should be >0.99).

See Fig. 2 for an example of a standard curve for *hdcA* gene.

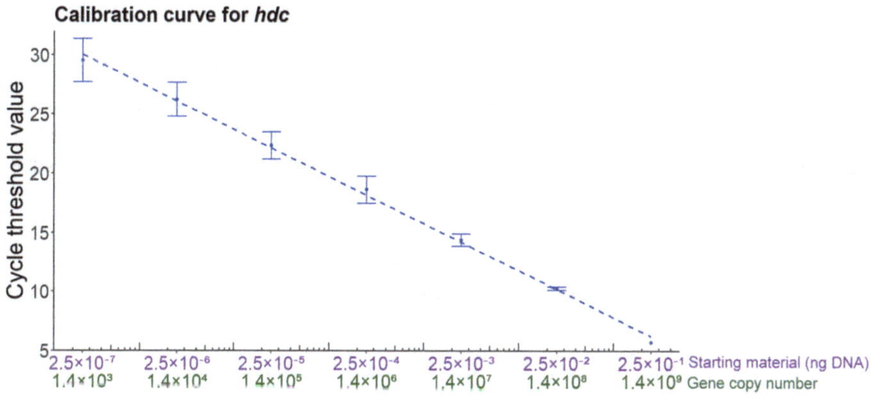

Fig. 2 Example of a standard curve depicting the Ct (cycle threshold) value obtained in qPCR experiments for each sequential dilution of the amplicon, expressed in ng of starting material (shown in purple) or in number of copies of the *hdcA* gene (shown in green). Resulting standard curve equation: $y = 2.77 - 3.82 \cdot \log(x)$, where x = starting material (ng DNA) and y = Ct value; $y = 40.0 - 3.82 \cdot \log(x)$, where x = gene copy number and y = Ct value

3.3 Development of a Standard Curve for a Marker Gene to Normalize Copy Number of the Target Gene

Normalization of gene copy number is a crucial step to accurately compare the bacterial load across samples and ensure consistency in the interpretation of the results. Typically, this involves dividing the target gene copy number by the copy number of a reference gene. In studies focused on bacterial samples, the 16S rRNA gene is generally used as the reference gene because of its ubiquitous presence in bacteria and high conservation across bacterial species.

To achieve this, follow the same procedure described in Step 3.1, with these distinctions:

- DNA extraction (Point 3.2.1): any bacterial species can be used in this stage, as the 16s rRNA gene is typically present in all bacterial cells in sufficient quantities. To obtain a representative sample of 16s rRNA copy numbers across different samples, the DNA can be extracted from bacterial grown on the surface of MRS plates following the GenElute™ Bacterial Genomic DNA Kit's instructions.

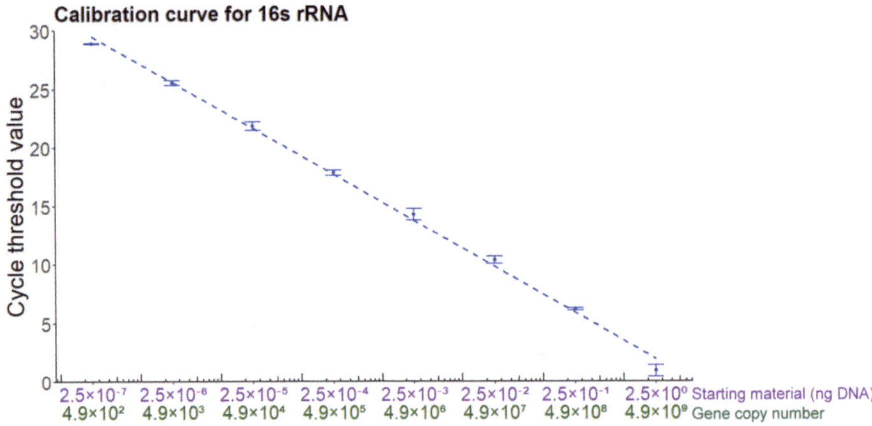

Fig. 3 Example of a standard curve depicting the Ct (cycle threshold) value obtained in qPCR experiments for each sequential dilution of the amplicon, expressed in ng of starting material (shown in purple) or in number of copies of the 16S rRNA gene (shown in green). Resulting standard curve equations: $y = 3.54 – 3.94 \cdot \log(x)$, where $x =$ starting material (ng DNA) and $y =$ Ct value; $y = 40.23 – 3.94 \cdot \log(x)$, where $x =$ gene copy number and $y =$ Ct value

- Primers used to amplify 16s rRNA (V3–V4): We recommend the primer pair:
- 5′-TCCTACGGGAGGCAGCAGT-3′ and 5′ - GGACTACCAGGGTATCTAATCCTGTT-3′ [12], which yields an amplicon of = 467 bp.

See Fig. 3 for a standard curve for the 16S rRNA gene.

3.4 Detection and Quantification of the Target Gene in Cheese Samples

3.4.1 Sample Cheese Collection and Homogenization

25 g of cheese is mixed with 225 mL of BPW in filter bags and homogenized at 300 revolutions per minute (rpm) during 5 min in a peristaltic homogenizer. In this way, a stock homogenate of each sample is obtained. The resulting stock homogenate is transferred to sterile bottles avoiding the cheese fat (*see* **Note 4**).

3.4.2 DNA Extraction of the Microbiota of Cheese

A 15-mL sterile Falcon is filled with 10 mL of the stock homogenate. This solution is centrifuged at 10.000 RCF for 20 min at 4°C (*see* **Note 5**). The supernatant is discarded, and without disturbing the pellet, the wall fat is removed using a sterile swab. The pellet is washed with 5 mL of BPW to remove as much of the fat matrix as possible. Then, an extra centrifugation is performed under the same conditions as described above. After discarding the supernatant and removing wall fat again, the pellet is resuspended using the first buffer of DNeasy™ PowerFood™ Microbial Kit, and the remaining steps are carried out according to the manufacturer's instructions (*see* **Note 6**). Finally, the DNA concentration of each sample is measured using the Qubit™ 4 fluorometer (or any other appropriate method) (*see* **Note 7**).

3.4.3 DNA Normalization and Conduction of qPCR

Before starting with the qPCR, it is important to normalize the DNA concentration. If the DNA concentration of the sample is higher than 25 ng/μL, dilute it with milli-Q sterile water as needed. Then, add 1 μL of the diluted solution into the qPCR well along with 7 μL of milli-Q sterile water. On the other hand, if the DNA concentration is lower than 25 ng/μL, adjust the volume of DNA added to ensure that each well contains 25 ng of DNA (1.25 ng/μL). Complete the remaining volume to reach 8 μL with milli-Q sterile water (*see* **Note 8**). Then, prepare a stock solution containing both primers and SYBR Green™ Master Mix. To do this, count the number of samples you will analyze for each gene (*see* **Note 9**). Finally, add the prepared mix to each qPCR well and ensure the solution reaches the bottom of the wells without creating bubbles, as bubbles can interfere with fluorescence detection.

It is recommended to analyze, for each DNA template, both *hdcA* and 16S rRNA genes in the same run. Additionally, it is essential to include both a positive and negative control for each primer pair to ensure that the qPCR has run correctly.

We also recommend including a melt curve at the end of the qPCR run to confirm the specificity and purity of the amplified products. A well-defined melt curve with a single peak at the expected melting temperature indicates that the qPCR results are reliable and that the quantification is based on the correct target gene. In contrast, if the melt curve displays signs of non-specific amplification, such as multiple peaks or a broad curve, the results should be regarded as unreliable or inconsistent.

3.5 Normalization of Gene Counts and Interpretation of Results

This section presents exemplary results illustrating the process of quantification and normalization of the non-marker gene after having obtained a standard curve for the non-marker gene (Fig. 2), a standard curve for the marker gene (Fig. 3), and Ct values for both genes from three DNA samples of cheese microbiota.

	Ct value	
Sample	***hdcA***	**16S rRNA**
1	24.25	7.89
2	22.12	11.34
3	16.79	10.02

To obtain the gene copy numbers for each sample, using the equations derived from the linear regressions in Figs. 2 and 3 (shown in green). Substitute the Ct values for y in each equation and solve for x.

Sample	Number of gene copies *hdcA*	16S rRNA
1	13,275	161,481,148
2	47,932	21,502,418
3	1,191,012	46,506,397

To compare the results, it is necessary to normalize the number of gene copies. This is done by dividing the *hdcA* copies by the 16S rRNA gene copies.

Sample	Normalized *hdcA* copy number
1	0.000082
2	0.002229
3	0.025610

The data show that the target gene, *hdcA*, is present in increasing amounts across the three samples, with Sample 3 showing the highest relative amount, followed by Sample 2 and Sample 1. Normalization against the 16S rRNA gene copies ensures that the observed differences in gene abundance are not attributable to variations in the total DNA quantity among the samples.

Overall, this methodology allows for a quick and efficient means to screen for BA-producing genes within cheese microbiota. It is recommended that only those samples displaying the highest relative abundances of the *hdcA* gene should be selected for BAs quantification using more costly methods (such as HPLC). This approach improves cost-effectiveness and enables high-throughput analysis of BA production as a screening method, reserving more resource-intensive techniques for cases that are more likely to pose food safety concerns.

4 Notes

1. During the final elution step, it is recommended to elute in 50 μL to achieve a good final DNA concentration.
2. It is recommended to yield a high DNA concentration (>25 ng/μL).
3. To ensure the accuracy and reliability of the standard curve, it is recommended to perform at least three technical replicates for each dilution point.

4. A filter bag equipped with a membrane is used to separate the liquid matrix from fat. It is crucial at this stage to ensure that only the liquid phase is collected, avoiding the fat matrix.
5. It is important to refrigerate the centrifuge previously and perform the process at low temperatures to reach an optimal separation of the fat layer.
6. To obtain an optimal final DNA concentration, the DNA is initially eluted in 25 μL of elution buffer. This same volume is then reapplied to the column for a second elution to maximize recovery.
7. It is recommended to store DNA at −20°C if experiments are planned for a different day.
8. The DNA concentration of each sample must be at least 3.125 ng/μL, as the maximum volume that can be added to the qPCR plate is 8 μL. This concentration ensures that a final DNA amount of 25 ng (1.25 ng/μL) per well can be achieved. For example, if a sample has a DNA concentration of 10 ng/μL, the volume of the sample to add can be calculated by dividing the desired final DNA amount by the initial concentration:

$$\frac{25ng}{10ng/\mu L} = 2.5\mu L$$

   Thus, add 2.5 μL of the sample to the well, and complete the remaining volume (to reach 8 μL) with milli-Q sterile water. This method ensures that each well has 25 ng of DNA at a final volume of 8 μL.
   An alternative, if the DNA concentration is too low to reach 3.125 ng/μL, is to evaporate under vacuum the water and resuspend the DNA in a smaller volume of milli-Q sterile water.
9. It is important to prepare a master mix containing all the reagents required for each gene. It is recommended to add the primers first, followed by the SYBR Green Master Mix. To minimize resource waste and improve reproducibility, it is advisable to calculate the number of samples to be analyzed and prepare a master mix with an extra volume to account for pipetting losses. A good practice is to prepare the mix by including an additional reaction to ensure sufficient volume.

References

1. Saha Turna N, Chung R, McIntyre L (2024) A review of biogenic amines in fermented foods: occurrence and health effects. Heliyon 10:e24501
2. Pawul-Gruba M, Kiljanek T, Madejska A, Osek J (2022) Development of a high performance liquid chromatography with diode array detector (HPLC-DAD) method for determination of

biogenic amines in ripened cheeses. Molecules 27:8194
3. Benkerroum N (2016) Biogenic amines in dairy products: origin, incidence, and control means. Compr Rev Food Sci Food Saf 15:801–826
4. Natrella G, Vacca M, Minervini F, Faccia M, De Angelis M (2024) A comprehensive review on the biogenic amines in cheeses: their origin, chemical characteristics, hazard and reduction strategies. Foods 13:2583
5. Wüthrich D, Berthoud H, Wechsler D, Eugster E, Irmler S, Bruggmann R (2017) The histidine decarboxylase gene cluster of *Lactobacillus parabuchneri* was gained by horizontal gene transfer and is mobile within the species. Front Microbiol 17:218
6. Wójcik W, Łukasiewicz M, Puppel K (2021) Biogenic amines: formation, action and toxicity – a review. J Sci Food Agric 101:2634–2640
7. Moniente M, García-Gonzalo D, Ontañón I, Pagán R, Botello-Morte L (2021) Histamine accumulation in dairy products: microbial causes, techniques for the detection of histamine-producing microbiota, and potential solutions. Compr Rev Food Sci Food Saf 20:1481–1523
8. Fernández M, del Río B, Linares DM, Martín MC, Alvarez MA (2006) Real-time polymerase chain reaction for quantitative detection of histamine-producing bacteria: use in cheese production. J Dairy Sci 89:3763–3769
9. Rossi F, Gardini F, Rizzotti L, La Gioia F, Tabanelli G, Torriani S (2011) Quantitative analysis of histidine decarboxylase gene (*hdcA*) transcription and histamine production by *Streptococcus thermophilus* PRI60 under conditions relevant to cheese making. Appl Environ Microbio l77:2817–2822
10. Bajrami D, Fischer S, Barth H, Sarquis MA, Ladero VM, Fernández M, Sportelli MC, Cioffi N, Kranz C, Mizaikoff B (2022) In situ monitoring of *Lentilactobacillus parabuchneri* biofilm formation via real-time infrared spectroscopy. NPJ Biofilms Microbiomes 8:92
11. Moniente M, Botello-Morte L, García-Gonzalo D, Virto R, Pagán R, Ferreira V, Ontañón I (2023) Combination of SPE and fluorescent detection of AQC-derivatives for the determination at sub-mg/L levels of biogenic amines in dairy products. Food Res Int 165:112448
12. Nadkarni MA, Martin FE, Jacques NA, Hunter N (2002) Determination of bacterial load by real-time PCR using a broad-range (universal) probe and primers set. Microbiology 148:257–266

Chapter 6

Selection and Genotypic Characterization of Potential Bacterial Strains for Use as Protective Cultures

Nerea Garín-Murguialday, Raquel Virto, Rafael Pagán and Laura Espina

Abstract

Antimicrobial compounds produced by bacteria, such as lactic acid bacteria and *Bacillus* spp., are alternatives to artificial preservatives, enabling companies to meet customer demand for clean-label products. To comply with the relevant EFSA's specifications, it is important to genotypically and phenotypically characterize the cultures that produce the antimicrobial compounds. This chapter summarizes the steps necessary to characterize the strains originating the protective cultures for their potential use as ingredients in the food industry.

Keywords Protective cultures, Cell-free supernatant, Antimicrobial activity, Bacteriocins, Antibiotic resistance, Whole genome sequencing

1 Introduction

Food on the market must be safe and innocuous, so preservation is a fundamental step to producing products with an adequate shelf life, suitable for consumption. Preservation must cope with two aspects. On the one hand, it should be controlling pathogenic microorganisms, such as *Salmonella* spp., *Escherichia coli,* and *Listeria monocytogenes*, which cause foodborne diseases [1]; and on the other hand, it should reduce food waste often caused by bacteria, molds (*Penicillium* spp., *Aspergillus* spp.), and yeasts [2, 3].

Companies are continuously working to develop a variety of techniques to inhibit microbial growth. Conventional technologies such as heat treatments (pasteurization and sterilization), cooling techniques (refrigeration and freezing), reduction in water activity (spray-drying and freeze-drying), fermentation, and chemical preservatives [4] have been instrumental in maintaining food safety. However, these processes do not fully align with consumers' desire for products that retain the organoleptic and nutritional characteristics of fresh products [5]. Consequently, the industry faces the dual

Marciane Magnani and Geany Targino de Souza Pedrosa (eds.), *Detection and Quantification of Bacteria, Yeast, Viruses, and Protozoan in Foods and Freshwater*, Methods and Protocols in Food Science, https://doi.org/10.1007/978-1-0716-5064-6_6,

challenge of extending product shelf life while aligning strategies with current consumer trends, such as the clean label movement [6].

In response to this consumers' demand, various novel preservation strategies have emerged, including the use of bioprotective bacterial cultures and their metabolites [7]. This approach involves utilizing microorganisms or their products, such as cell-free supernatants (CFSs), that act through competitive exclusion or the production of antimicrobial substances [8]. Following European Commission's requirements on the characterization of production microorganisms used in feed [9, 10], the selected microbial species must belong to the Qualified Presumption of Safety (QPS) [11] and must not possess any acquired transferable antibiotic resistance genes to prevent horizontal gene transfer. Furthermore, there must be evidence of the non-toxigenicity of the cultures, the stability of the intended technological effect, and the definition of safe usage levels [12, 13]. Among the possible options, lactic acid bacteria (LAB) stand out due to their well-established tradition in food applications, their perception as safe for human consumption, and the exhibition of a broad spectrum of antimicrobial activity (AM) [12]. Another interesting potential protective culture is based on the exploitation of some *Bacillus* spp., which produce a wide range of antimicrobial and antifungal compounds [14].

In this chapter, we describe a methodology for the comprehensive evaluation of potential protective cultures and the verification of their compliance with all the relevant EFSA's specifications, combining phenotypic assays and genome analysis of sequencing data. More specifically, we summarize how to measure the antimicrobial activity by agar well diffusion assay and by minimum inhibitory concentration (MIC) determination, assess antibiotic resistance and cytotoxicity, infer the nature of the antimicrobial compounds and their thermal stability, and analyze genomic data for taxonomic confirmation and identification of antibiotic resistance genes (ARGs) and biosynthetic gene clusters (BGCs) involved in secondary metabolite production. We have validated this protocol for lactic acid bacteria and *Bacillus subtilis*, but it can be applied for any potential protective culture. In addition, we offer tips to show you how to prepare the material, follow the methods, and interpret the results.

2 Materials

The material required to perform the characterization of the microbial strains is the following:

2.1 Culture Preparation

1. Cryovial of the indicator microorganism.
2. Cryovial of the potential protective culture microorganism.
3. Broth selected for optimal growth of each microorganism.

4. Sterile bottles of 10 and 50 mL.
5. Agar selected for optimal growth of each microorganism.
6. Sterile petri dishes (90 mm).
7. Micropipettes.
8. Sterile pipette tips.
9. Microtubes (1.5 mL).
10. Buffered peptone water (0.1%).
11. Incubator.
12. Centrifuge.
13. Single-use 500-mL vacuum filter unit with 0.22-μm PES filter.
14. Sterile plastic bottles.

2.2 Measurement of Antimicrobial Activity of the CFS Using the Agar Well Diffusion Method

1. Sterile petri dishes (90 mm).
2. Optimum growth agar of the indicator microorganism.
3. Sterile borer (7.20 mm).
4. Calliper or ruler.
5. Buffered peptone water (0.1%).
6. Digralsky seeding loops.
7. Sterile petri dishes (90 mm).
8. Optical microscope.
9. Incubator.

2.3 Determination of the MIC of the CFS Using the Broth Microdilution Assay

1. 96-well flat-bottom microtiter plate.
2. Micropipettes.
3. Sterile pipette tips.
4. Microtubes (1.5 mL).
5. Optimum growth broth of the indicator microorganism.
6. Sterile bottles of 10 and 50 mL.
7. Spectrophotometer.
8. Incubator.

2.4 Measurement of the Antibiotic Resistance of the Protective Cultures Using the Broth Microdilution Assay

1. Sterile petri dishes (90 mm).
2. Anaerobiosis cabinet.
3. Bunsen burner.
4. 96-well flat-bottom microtiter plate.
5. Multichannel pipette.
6. Micropipettes and sterile pipette tips.
7. Pipettor and sterile serological pipettes (10 and 25 mL).

8. Sterile microtubes (1.5 and 2 mL).
9. Sterile filters (0.22 µm).
10. Sterile falcon (15 and 20 mL).
11. Antibiotics (kanamycin, gentamicin, streptomycin, tetracycline, clindamycin, chloramphenicol, ampicillin, erythromycin, and vancomycin).
12. Incubators.
13. Iso-Sensitest (IST) broth.
14. Man Rogose y Sharpe Medium (MRS broth and agar).
15. Screw-capped glass culture tubes.
16. UV/VIS spectrophotometer.
17. Spectrophotometer cuvettes.
18. Anaerobiosis jar.
19. Anaerocult® for microbiology.
20. Sterile NaCl 0.85%.
21. MiliQ sterile water.

2.5 Measurement of Cytotoxicity of the CFS Using the MTT Assay

1. Caco-2 cell line.
2. HT-29 cell line.
3. 3-(4, 5-dimethylthiazol-2-yl)-2, 5-diphenyltetrazolium bromide.
4. 96-well flat-bottom microtiter plate.
5. Micropipettes.
6. Sterile pipette tips.
7. Incubator CO_2 control.
8. Freezer (−80°C).
9. Phenol red-free medium.
10. Dimethyl sulfoxide (DMSO).
11. Spectrophotometer.

2.6 Determination of the Nature of the Antimicrobial Compounds of the CFS

1. NaOH 1 M.
2. HCl 1 M.
3. Proteinase K.
4. Phosphate.
5. Pepsin.
6. Sterile pasteur pipettes.
7. Incubator.
8. Micropipetes.

9. Sterile pipette tips.
10. Optimum growth agar of the indicator microorganism.
11. Digralsky seeding loops.
12. L-shaped spreaders.
13. Sterile petri dishes (90 mm).
14. Sterile borer (7.20 mm).
15. Caliper or ruler.
16. Buffered peptone water (0.1%).
17. Microtubes (1.5 mL).

2.7 Determination of the Thermal Stability of the CFS

1. Sterile micropipettes.
2. Sterile pipette tips.
3. Vials (2 mL).
4. Microtubes (2 mL).
5. Box with cold water or ice.
6. Thermoblock.
7. Calibration oil bath.
8. Optimum growth agar of the indicator microorganism.
9. Digralsky seeding loops.
10. L-shaped spreaders.
11. Sterile petri dishes (90 mm).

2.8 DNA Extraction, Whole Genome Sequencing (WGS) and Post-Sequencing Analysis of Each Potential Culture for the Determination of ARGs and of BGCs

2.8.1 DNA Extraction

1. Pure culture of the protective strain.
2. DNA extraction kit (GenElute Bacterial Genomic DNA kit, NA2110, Sigma-Aldrich, Missouri, USA).
3. Microcentrifuge.
4. Micropipettes.
5. Microtubes.
6. Pipette tips.

2.8.2 WGS: Illumina-Based Whole Genome Sequencing Service

Post-Sequencing Analysis

1. Paired-end raw FASTQ files with short reads generated from the submitted DNA.
2. Bioinformatic tools and their dependencies accessed on a local Linux server or via their standalone GUI application, or alternatively using the European Galaxy Server [15]:

   a. Trimmomatic v0.39 [16].
   b. FastQC v0.12.1 [17].
   c. Kraken2 v2.1.3 against a comprehensive and updated database [18].
   d. Shovill v1.1.0 [19].
   e. QUAST v5.2.0 [20].
   f. ABRicate v1.0.1 [21] with CARD [22, 23] and Resfinder [24, 25] databases.
   g. AntiSMASH v6.1.1 [26].
   h. BAGEL4 [27].

3 Methods

3.1 Culture Preparation

The first step in the analysis is the preparation of the cultures of the indicator microorganisms [28–30] and the potential protective cultures. The selection of the indicator microorganisms, whether pathogenic or spoilage, should be guided by their associated effects and prevalence in the targeted food products. Some potential indicators include *Escherichia coli*, *Listeria monocytogenes*, and *Salmonella* spp. [1]. For both indicator microorganisms and protective cultures, it is essential to conduct a preliminary study to determine the optimal growth conditions for each bacterial strain, including the choice of broth and agar media, incubation temperature, atmospheric conditions, and growth time.

3.1.1 Preparation of Cultures of the Indicator Microorganisms

Inoculate and spread the indicator strain from the cryovial on the surface of agar plates with an inoculation loop to isolate colonies and then, incubate the plates under the indicator microorganism's optimal growth conditions. Select 2 colonies of the plate, inoculate them into 10 mL of broth in a glass bottle with caps, and incubate the inoculated bottles under the optimal growth conditions. After that time, inoculate 50 mL of growth broth in a glass bottle with 100 μL of the previous culture and incubate the inoculated bottle (*see* **Note 1**).

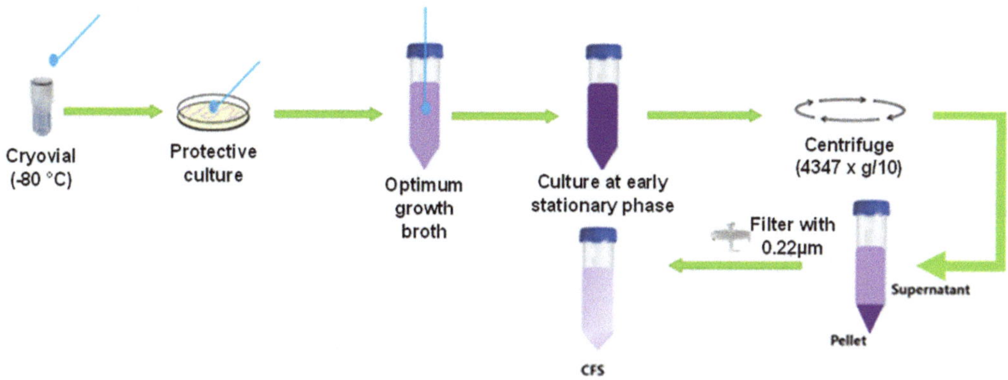

Fig. 1 Methodology to obtain the CFS of the protective cultures

3.1.2 Preparation of Protective Cultures and Collection of the Cell-Free Supernatants (CFS)

Figure 1 shows how to prepare the protective culture's CFS. Select a potential protective culture strain from the QPS list drawn up by EFSA and spread it on the surface of agar plates selected for their optimal growth with an inoculation loop. Incubate the plate under the protective microorganism's optimal growth conditions. Inoculate 3–4 colonies of it into 50 mL of growth broth in a glass bottle with caps and incubate them. Centrifuge 50 mL of the culture at 4347 × *g* for 10 min at 4°C to obtain the supernatant. Filter it with a 0.22 µm filter and store the CFSs at 4°C.

3.2 Measurement of Antimicrobial Activity of the CFS Using the Agar Well Diffusion Method

Pour 20 mL of the agar selected for the optimal growth of the indicator strain and let it solidify. Then, add and spread 10^6 CFU/mL of the indicator strain with a Digralsky loop and punch a well of 7.2 mm in diameter in each plate with the agar using a sterile borer. Fill each well with 100 µL of the testing CFS. Finally, keep the plates for 2 h on the bench and then incubate them at the optimal growth conditions of the indicator strain as indicated in Fig. 2.

Measure the halo the way Fig. 3 shows and select the strains with the highest antimicrobial activity (*see* **Note 2**).

3.3 Determination of the MIC of the CFS Using the Broth Microdilution Assay

Add to each well of a sterile 96-well flat-bottom microtiter plate 180 µL of the selected broth for the optimal growth broth of the indicator strain. Fill the wells to reach variable concentrations of the CFSs (0%, 0.15%, 0.31%, 0.625%, 1.25%, 2.5%, and 5%) and with the indicator microorganism at a final concentration of 10^3 CFU/mL and 10^5 CFU/mL. Incubate the plates under optimum growth conditions for the indicator microorganism and measure the optical density at a wavelength of 620 nm at different times (*see* **Note 3**). With the obtained results determine the MIC as the lowest concentration of a compound necessary to inhibit bacterial growth.

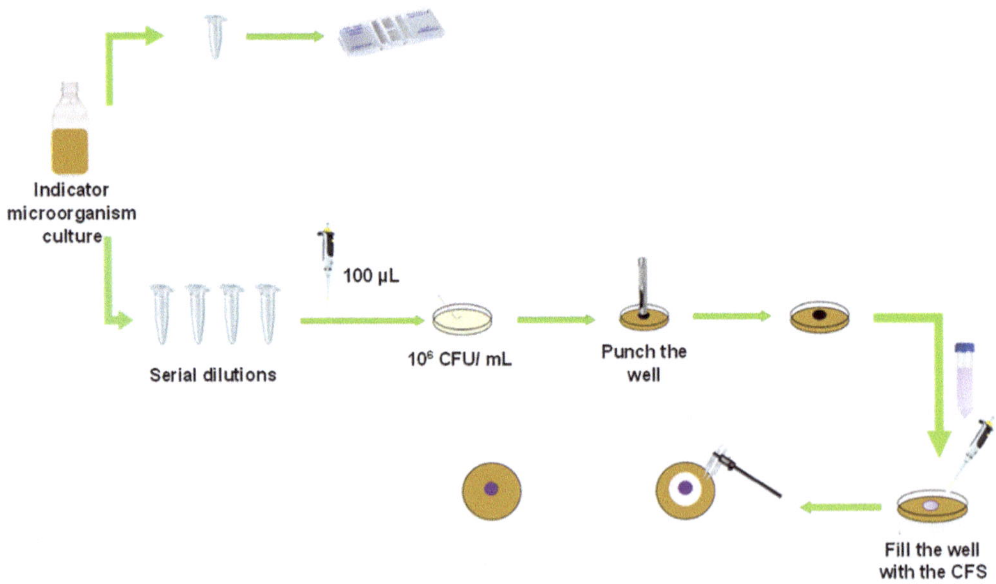

Fig. 2 Agar well diffusion assay

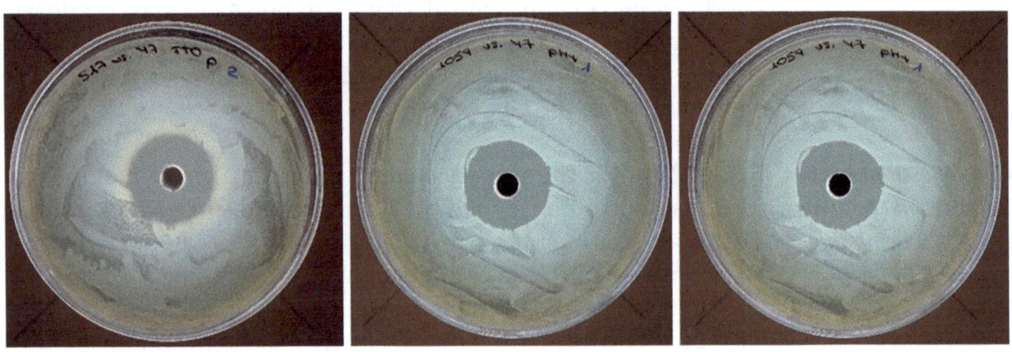

Fig. 3 Halo measurement

3.4 Measurement of the Antibiotic Resistance of the Protective Cultures via the Broth Microdilution Assay

The determination of antibiotic resistance of the strains is carried out phenotypically using the modified ISO 10932/IDF 223 standard [31]. First, seed the bacteria to be tested and incubate them in their optimal growth conditions by streaking by depletion and in parallel prepare the 0.85% NaCl, MRS Broth, IST Broth and miliQ water solutions and the antibiotic stocks and store them at −20°C. Add the culture medium to the wells of the microtiter plate and 10 μL of each of the antibiotics in column 2 and in the rows indicated in the ISO. Resuspend in 5 mL of the saline solution (NaCl 0.85%) the amount of colonies of the bacteria to be tested to an OD nm of 0.16–0.2 (3×10^8 CFU/mL). Dilute the above suspensions at 1/500 in the appropriate culture medium and distribute 50 μL of the 1/500 diluted bacterial suspension from column 2 to column 12 per well. To finish, incubate at the appropriate temperature and conditions for 48 h.

3.5 Measurement of Cytotoxicity of the CFS Using the MTT Assay

The in vitro assessments are conducted using Caco-2 and HT-29 cell lines, both human colorectal cancer cells, by MTT method (3-(4,5-dimethylthiazol-2-yl)-2,5-diphenyltetrazolium bromide) to determine if the strains to be used as ingredients prove cytotoxic to human health [32]. Seed the cell lines in a 96-well flat-bottom microtiter plate (5×10^3 cells/well for the Caco-2 and 1.7×10^4 cells/well for the HT-29 cell line) and then incubate them at 37°C (48 h, 5% CO_2, and 95% relative humidity). For the HT-29 cell line, directly add the CFS at different concentrations. For the Caco-2 cell lines, refresh the medium three times per week and after 15 days introduce the CFS at varying concentrations. Transfer the entire volume to 96-well-U-bottom microtiter plates and store at −80°C. Afterwards, remove the culture medium and add to each well 200 μL of phenol red-free medium and 50 μL of MTT (3-(4,5-dimethylthiazol-2-yl)-2,5-diphenyltetrazolium bromide) at a concentration of 5 mg/mL. Incubate the plates at 37°C (3 h, 5% CO_2, and 95% relative humidity) and eliminate the unmetabolized MTT. When eliminated, add 150 μL/well of dimethyl sulfoxide (DMSO) and agitate at 347.76 × *g* for 5 min. Transfer 100 μL of the supernatant to a new 96-well flat-bottom plate to measure the optical density at a wavelength of 570 nm. Finally, calculate the viability of the cell lines using Eq. (1).

$$\%\ viability = \frac{OD570\ sample}{OD570\ control} \times 100 \qquad (1)$$

3.6 Determination of the Nature of the Antimicrobial Compounds of the CFS

To ensure that the antimicrobial activity of CFS is due to the production of bacteriocins by the strain, neutralization (1.a), acidification (1.b) or addition of proteases (1.c–1.d) to the CFS is carried out.

1.a. Add 1 M NaOH to the CFS (pH 7.00) [33, 34].
1.b. Add 1 M HCl to the CFS (pH 3.50) [33, 34].
1.c. 25 μL of proteinase K to 5 mL of CFS [35].

Perform measurement of antimicrobial activity of the CFS using the agar well diffusion method (3.2) and measure the halos and determine whether the activity is due to the organic acids or to the production of bacteriocins (*see* **Note 4**).

3.7 Determination of the Thermal Stability of the CFS

The heat resistance of the CFSs is evaluated to determine whether their antimicrobial activity would remain effective when incorporated into foods prior to pasteurization or sterilization treatments commonly used in the food industry. These treatments were used as a reference to select the intensity (temperature and time) of the different heat treatments (from 10 to 120 min at 70, 90, 110, and 120°C).

To perform it, add a small volume (1 mL) of the CFS to a vial with 9 mL of preheated treatment medium (e.g., culture media, buffer, liquid food, etc.) and introduce it into a preheated liquid at the selected temperature (*see* **Note 5**). After the established time remove the vials and place them in cold water for thermal shock. Repeat the measurement of the antimicrobial activity of the CFS using the agar well diffusion method (3.2) with the treated CFS. Measure the halo obtained and compare it with the original halo made with the CFS without treatments (*see* **Note 6**) using Eq. (2).

$$\%\ \textit{loss antimicrobial activity} = \frac{\textit{halo diameter (mm) before heat treatment} - \textit{halo diameter (mm) after heat treatment}}{\textit{halo diameter (mm) before treatment}} \times 100 \quad (2)$$

3.8 DNA Extraction, WSG and Post-Sequencing Analysis of Each Potential Culture for the Determination of ARGs and BGCs

Besides carrying out the aforementioned experimental procedures for the determination of the antibiotic resistance and of the potential nature of the antimicrobial compounds, WGS and post-sequencing analysis should also be carried out to help reveal the possible ARGs and the potential BGCs involved in secondary metabolite production, respectively.

3.8.1 DNA Extraction

A pure culture of the potential protective strain must undergo DNA extraction with an adequate DNA extraction kit, following its instructions.

3.8.2 WGS

A sequencing provider can perform DNA quality checking, library preparation and sequencing on an Illumina platform, delivering high-quality sequencing reads with Phred scores above 30 for most bases.

3.8.3 Post-Sequencing Analysis

The resulting paired-end raw FASTQ files are then processed following the steps showed in Fig. 4. First, pre-process the 2 × 150-bp paired-end sequencing reads generated by removing adapters and short reads using Trimmomatic v0.39 with default settings for sliding window quality control, leading/trailing base quality trimming, and a minimum length of 36 bp and check the quality of the filtered clean reads with FastQC v0.12.1. Then use Kraken2 on the raw FASTQ reads to classify sequences against a comprehensive database (e.g., bacterial, archaeal, and viral) to confirm the taxonomic identity of the bacterial isolate and the absence of contaminants and perform the novo genome assembly with Shovill v.1.1.0 using the SPAdes assembler with default parameters. Evaluate the quality of the assembly with QUAST v5.2.0. and, as an optional step, run Kraken2 again on the FASTA file for additional confirmation of the taxonomic identity of the assembled contigs. Screen the assembled contigs in FASTA format for ARGs using the ABRicate v1.0.1 tool with the CARD and ResFinder databases with a minimum identify

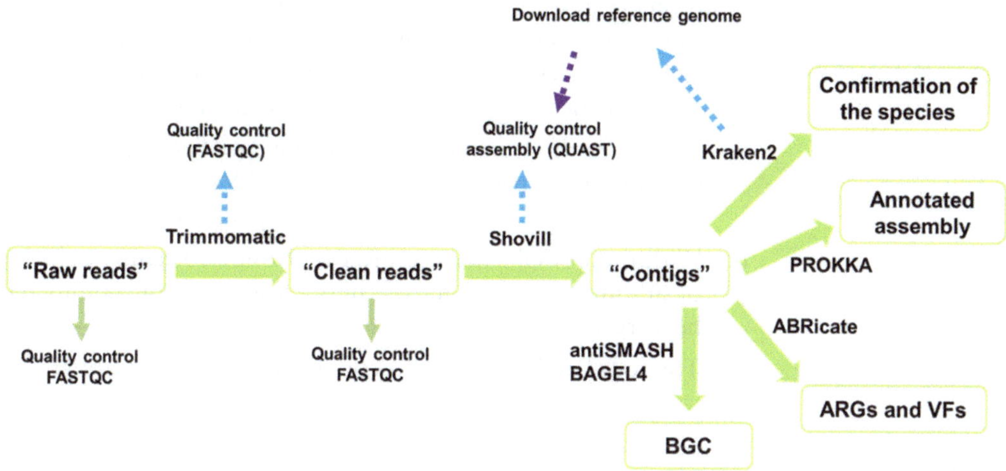

Fig. 4 Bioinformatic analysis of the WGS of the antimicrobial strain

and coverage values of 80%. Finally, predict the type and location of BGCs involved in secondary metabolite biosynthesis within the assembled genomes with antiSMASH v6.1.1 (*see* **Note 7**).

Final remarks: The use of Kraken2 ensures compliance with the European Commission's requirements for accurate identification of cultures. The identification of ARGs enhances the information provided by the phenotypic tests detailed in Sect. 3.4. The identification of BGCs via antiSMASH complements the data obtained from the determination of the nature of the antimicrobial compound present in the CFS, as described in Step 3.6.

4 Notes

1. To verify the efficacy of the filtration process, streak the obtained CFS and incubate it under the growth conditions of the protective bacteria. If the filtration is inadequate, repeat the filtration process.
2. When quantifying the halos, it must be considered that not all halos are positive, but only those exceeding the diameter of the well (7.2 mm) by 2 mm (>9.2 mm).
3. The MIC can be determined in two ways: by measuring at different intervals after removing the plate from the incubator, or by using an incubator that continuously measures while maintaining incubation conditions, thereby obtaining the growth curves of the indicator microorganism in the presence of various concentrations of the CFS.
4. If the halo disappears when neutralizing the CFS, the antimicrobial activity can be attributed to the organic acids produced by the bacteria. If the halo disappears after the addition of the

protease, the antimicrobial activity of the CFS can be ascribed to the production of proteinaceous bacteriocin or bacteriocin-like substances. However, if the halo persisted, other proteases, such as pancreatin, should be added to the tests.

5. To ensure the treatment starts at the specified time, the temperature of a separate vial containing CFS under identical conditions must be monitored with a temperature probe.
6. With these results make a linear regression to estimate the loss of antimicrobial activity of the CFS at different commercial heat treatments.
7. The tool BAGEL4 is also recommended for this purpose.

References

1. Elika (2023) Informe RASFF alimentos alertas y notificaciones. https://alimentos.elika.eus/wp-content/uploads/sites/2/2024/02/resumen-anual-_rasff-alimentos_2023.pdf (accessed on 13 October 2024)
2. Strack L, Carli RC, Silva RVd, Sartor KB, Colla LM, Reinehr CO (2020) Food biopreservation using antimicrobials produced by lactic acid bacteria. Res Soc Dev 9:e998986666
3. Rawat S (2015) Food spoilage: microorganisms and their prevention. Asian J Plant Sci Res 5:47–56
4. Sridhar A, Ponnuchamy M, Kumar PS, Kapoor A (2021) Food preservation techniques and nanotechnology for increased shelf life of fruits, vegetables, beverages and spices: a review. Environ Chem Lett 19:1715–1735
5. Delgado-Pando G, Ekonomou SI, Stratakos AC, Pintado T (2021) Clean label alternatives in meat products. Foods 10:71615
6. Aschemann-Witzel J, Varela P, Peschel AO (2019) Consumers' categorization of food ingredients: do consumers perceive them as 'clean label' producers expect? An exploration with projective mapping. Food Qual Prefer 71:117–128
7. Barcenilla C, Ducic M, Lopez M, Prieto M, Alvarez-Ordonez A (2022) Application of lactic acid bacteria for the biopreservation of meat products: a systematic review. Meat Sci 183:108661
8. Hartmann HA, Wilke T, Erdmann R (2011) Efficacy of bacteriocin-containing cell-free culture supernatants from lactic acid bacteria to control *Listeria monocytogenes* in food. Int J Food Microbiol 146:192–199
9. EFSA (2005) Opinion of the Scientific Committee on a request from EFSA related to a generic approach to the safety assessment by EFSA of microorganisms used in food/feed and the production of food/feed additives. EFSA J 3:226
10. EFSA (2018) Panel on Additives and Products or Substances used in Animal Feed (FEEDAP); Rychen G, Azimonti G, Bampidis V et al Guidance on the characterisation of microorganisms used as feed additives or as production organisms. EFSA J 16:e05206
11. EFSA BIOHAZ Panel (2014) Updated list of QPS-recommended microorganisms for safety risk assessments carried out by EFSA. https://zenodo.org/records/10534041. Accessed 25 Apr 2024
12. Souza LV, Martins E, Moreira I, de Carvalho AF (2022) Strategies for the development of bioprotective cultures in food preservation. Int J Food Microbiol 2022:6264170
13. Ibrahim SA, Ayivi RD, Zimmerman T, Siddiqui SA, Altemimi AB, Fidan H, Esatbeyoglu T, Bakhshayesh RV (2021) Lactic acid bacteria as antimicrobial agents: food safety and microbial food spoilage prevention. Foods 10:123131
14. Qin Y, Wang Y, He Y, Zhang Y, She Q, Chai Y, Li P, Shang Q (2019) Characterization of Subtilin L-Q11, a novel Class I bacteriocin synthesized by *Bacillus subtilis* L-Q11 isolated from orchard soil. Front Microbiol 10:484
15. The Galaxy Community (2024) The Galaxy platform for accessible, reproducible, and collaborative data analyses: 2024 update. Nucleic Acids Res 52:W83–W94
16. Bolger AM, Lohse M, Usadel B (2024) Trimmomatic: a flexible trimmer for Illumina sequence data. Bioinformatics 30:2114–2120
17. Andrews S (2010) FastQC: a quality control tool for high throughput sequence https://www.bioinformatics.babraham.ac.uk/projects/fastqc/. Accessed 2 Nov 2024
18. Wood DE, Salzberg SL (2014) Kraken: ultrafast metagenomic sequence classification using exact alignments. Genome Biol 15:R46

19. Seeman T (2017) Shovill: faster SPAdes assembly of illumina reads. https://github.com/tseemann/shovill. Accessed 2 Nov 2024
20. Mikheenko A, Prjibelski A, Saveliev V, Antipov D, Gurevich A (2018) Versatile genome assembly evaluation with QUAST-LG. Bioinformatics 34:i142–i150
21. Seeman T (2016) ABRicate: mass screening of contigs for antibiotic resistance genes. https://github.com/tseemann/abricate. Accessed 2 Nov 2024
22. Alcock BP, Huynh W, Chalil R, Smith KW, Raphenya AR, Wlodarski MA, Edalatmand A, Petkau A, Syed SA, Tsang KK, Baker SJC, Dave M, McCarthy MC, Mukiri KM, Nasir JA, Golbon B, Imtiaz H, Jiang X, Kaur K, Kwong M, Liang ZC, Niu KC, Shan P, Yang JYJ, Gray KL, Hoad GR, Jia B, Bhando T, Carfrae LA, Farha MA, French S, Gordzevich R, Rachwalski K, Tu MM, Bordeleau E, Dooley D, Griffiths E, Zubyk HL, Brown ED, Maguire F, Beiko RG, Hsiao WWL, Brinkman FSL, Van Domselaar G, McArthur AG (2023) CARD 2023: expanded curation, support for machine learning, and resistome prediction at the Comprehensive Antibiotic Resistance Database. Nucleic Acids Res 51:D690–D699
23. Jia B, Raphenya AR, Alcock B, Waglechner N, Guo P, Tsang KK, Lago BA, Dave BM, Pereira S, Sharma AN, Doshi S, Courtot M, Lo R, Williams LE, Frye JG, Elsayegh T, Sardar D, Westman EL, Pawlowski AC, Johnson TA, Brinkman FS, Wright GD, McArthur AG (2017) CARD 2017: expansion and model-centric curation of the comprehensive antibiotic resistance database. Nucleic Acids Res 45:D566–D573
24. Florensa AF, Kaas RS, Clausen P, Aytan-Aktug D, Aarestrup FM (2022) ResFinder: an open online resource for identification of antimicrobial resistance genes in next-generation sequencing data and prediction of phenotypes from genotypes. Microb Genom 8:000748
25. Zankari E, Hasman H, Cosentino S, Vestergaard M, Rasmussen S, Lund O, Aarestrup FM, Larsen MV (2012) Identification of acquired antimicrobial resistance genes. J Antimicrob Chemother 67:2640–2644
26. Blin K, Shaw S, Kloosterman AM, Charlop-Powers Z, van Wezel GP, Medema MH, Weber T (2021) AntiSMASH 6.0: improving cluster detection and comparison capabilities. Nucleic Acids Res 49:W29–W35
27. van Heel AJ, de Jong A, Song C, Viel JH, Kok J, Kuipers OP (2018) BAGEL4: a user-friendly web server to thoroughly mine RiPPs and bacteriocins. Nucleic Acids Res 46:W278–W281
28. Lestari SDSA, Safitri WA, Dewi SS, Prastiyanto ME (2019) The antibacterial activities of bacteriocin *Pediococcus acidilactici* of breast milk isolate to against methicillin-resistant *Staphylococcus aureus*. J Phys Conf Ser 1374:012021
29. Ivanovic MM, Mirkovic M, Miocinovic J, Radulovic A, Knudsen TS, Radulovic Z (2021) Autochthonous *Enterococcus durans* pfmi565 and *Lactococcus lactis* subsp. *lactis* bgbu1-4 in bio-control of *Listeria monocytogenes* in ultrafiltered cheese. Foods 10:1448
30. Iseppi RC, Sabia C, Messi P (2020) Combined antimicrobial use of essential oils and bacteriocin bacLP17 as seafood biopreservative to control *Listeria monocytogenes* both in planktonic and sessile forms. Res Microbiol 171:351–356
31. International Organization for Standardization (2010) ISO 10932:2010 Milk and milk products - determination of the minimal inhibitory concentration (MIC) of antibiotics applicable to bifidobacteria and non-enterococcal lactic acid bacteria (LAB) https://www.iso.org/standard/56479.html
32. Mosmann T (1983) Rapid colorimetric assay for cellular growth and survival: application to proliferation and cytotoxicity assays. J Immunol Methods 65:55–63
33. Lin TH, Pan TM (2019) Characterization of an antimicrobial substance produced by *Lactobacillus plantarum* NTU 102. J Microbiol Immunol Infect 52:409–417
34. Selvam D, Thangarasu A, Shyu DJH, Neelamegam R, Muthukalingan K, Nagarajan K (2020) Antimicrobial substance produced by *Pseudomonas aeruginosa* isolated from slaughterhouse sediment: physicochemical characterization, purification, and identification. Int J Pept Res Ther 27:887–897
35. Darbandi A, Asadi A, Mahdizade Ari M, Ohadi E, Talebi M, Halaj Zadeh M, Emamie AD, Ghanavati R, Kakanj M (2022) Bacteriocins: properties and potential use as antimicrobials. J Clin Lab Anal 36:e24093

Chapter 7

Phenotypic Characterization of Antimicrobial-Resistant Variants in Foodborne Pathogens

Alberto Fau, Jorge Andaluz-Arbe, Rafael Pagán, Diego Gómez-Lozano, Diego García-Gonzalo and Daniel Berdejo

Abstract

The food industry uses multiple antimicrobial agents to control and eliminate spoilage and pathogenic bacteria throughout the food chain. However, the selective pressure placed by these compounds can lead to the development of resistances that spread from farm to fork. This resistance emerges through horizontal gene transfer (HGT) or genetic adaptation, leading to increased tolerance and/or resistance to antimicrobial agents, such as antibiotics, biocides, and essential oils. The increase in tolerance and/or resistance could compromise the efficacy of the antimicrobial agents and affect food safety. To phenotypically characterize study strains and detect genetically resistant variants (RVs), this chapter provides detailed protocols for determining the minimum inhibitory concentration (MIC), minimum bactericidal concentration (MBC), growth kinetics, and survival curves against antimicrobial compounds. These methods enable comparison between parental strain and study strains, allowing for precise assessment of growth fitness, direct- and cross-tolerance and/or -resistance levels across different agents and treatments.

Keywords Phenotypic characterization, Antimicrobial agent, Antimicrobial resistance (AMR), Genetically resistant variants (RVs), Direct- and cross-tolerance and/or resistance

1 Introduction

Foodborne diseases are a significant threat to food safety. In 2023, the European Food Safety Authority (EFSA) [1] reported more than 5500 foodborne outbreaks in Europe. Although there are several agents involved in foodborne outbreaks, pathogenic bacteria remain the main cause. Specifically, *Campylobacter* spp., *Salmonella* spp., *Shiga toxin-producing Escherichia coli* (STEC), and *Listeria monocytogenes* have been the most reported pathogen microorganisms over the last years. To control pathogenic bacteria, the food industry applies multiple strategies like the use of antimicrobial agents (biocides, food preservatives, essential oils, etc.) or food preservation methods (heat treatment, high hydrostatic pressures, ionizing radiation, etc.).

Marciane Magnani and Geany Targino de Souza Pedrosa (eds.), *Detection and Quantification of Bacteria, Yeast, Viruses, and Protozoan in Foods and Freshwater*, Methods and Protocols in Food Science, https://doi.org/10.1007/978-1-0716-5064-6_7,

The foodborne disease problem is exacerbated by the development of antimicrobial resistance (AMR). The emergence of AMR is conducted by two main strategies [2] whereby microorganisms acquire DNA with resistance genes through horizontal gene transfer (HGT) or through the emergence of random mutations that confer protection. These mutations can confer increased tolerance when the bacterial population is able to survive a transient exposure to the antimicrobial agent; or increased resistance, when the population is able to grow in the presence of the antimicrobial agent [3].

Antibiotic resistance is widely studied in several microorganisms related to food-borne diseases [4–7], this resistance hampers therapeutic treatments and has a significant impact on public health. However, this resistance phenomenon is also being observed with other antimicrobial agents like biocides [6, 8, 9], or essential oils [10, 11]. Exposure to an antimicrobial agent can lead to an increase in direct-tolerance and/or -resistance, but can also lead to cross-tolerance and/or -resistance, which is an increase in protection to external agents to which no previous exposure has occurred [12]. This phenomenon has also been reported for other antimicrobial products, such as antibiotics, biocides, and essential oils [8, 10, 13, 14], but also against other stress agents typically used in the food industry as preservation methods: acidification, osmotic stress, heat treatments, high hydrostatic pressures, etc. [11, 15–17]. As a consequence, the food chain might act as a vehicle for the spreading of AMR from farm to fork [2, 4, 18]. The increase in resistance could affect the efficacy of treatments applied in the food industry and threaten food safety.

Adaptive laboratory evolution assays (ALEs) are used to study the occurrence of these random mutations in vitro [19]. Three main protocols are available for this purpose. The first protocol involves applying constant sub-inhibitory doses of the antimicrobial during bacterial growth of a parental strain [20]. The second protocol consists of the application of increasing doses of the antimicrobial during bacterial growth of parental strain [21]. In the third main protocol, lethal treatments are applied followed by a recovery phase and growth of the surviving bacteria, which are the most tolerant in the population [20]. The isolation of evolved strains from ALEs and the comparison of their tolerance/resistance to that of the parental strain might allow for the detection of genetically resistant variants (RVs).

Derived customized variants can also be generated using genetic engineering techniques. Two main protocols are available for this purpose. The first protocol involves introducing plasmids into the bacteria to acquire resistance genes without integrating into the chromosome [22]. The second protocol involves replacing specific chromosomal DNA fragments with mutated sequences to stably incorporate resistance genes [23]. Transformed bacteria with antimicrobial resistance genes allow to assess the influence of specific antimicrobial resistance genes on the development of antimicrobial resistance.

This chapter aims to describe the methodology needed to phenotypically characterize strains obtained through ALEs from a parental strain or custom strains generated by insertion of extrachromosomal plasmids or gene edition by linear DNA integration. The objective is to evaluate the emergence of direct-tolerance and/or -resistance against the antimicrobial agent and cross-tolerance and/or -resistance against other antimicrobial agents, including food preservation treatments. For this purpose, different methodologies will be described to determine minimum inhibitory concentration (MIC), minimum bactericidal concentration (MBC), growth kinetics, and survival curves against antimicrobial compounds.

2 Material

2.1 Preparation of the Bacterial Working Culture

1. Cryovial of the bacterial strains to be studied.
2. Culturing tools: micropipettes, pipette tips, plastic 1.5-mL tubes, Petri dishes (90 mm), inoculation loops, and L-shaped spreaders.
3. Laboratory weighing material: precision balance, spoon, and spatula.
4. Growth media: tryptic soya broth and agar supplemented with 0.6% yeast extract (or any other nutritive broth or agar).
5. Glass test tubes and caps.
6. Sterile glass bottle narrow neck and screw caps.

2.2 Determination of the Minimum Inhibitory Concentration (MIC)

1. Bacterial cultures to be studied.
2. Culturing tools: micropipettes, pipette tips, plastic 1.5-mL tubes, and multichannel micropipette.
3. Laboratory weighing material: precision balance, spoon, and spatula.
4. Growth media: tryptic soya broth and agar supplemented with 0.6% yeast extract (or any other nutritive broth or agar).
5. Antimicrobial agents (food preservatives, biocides, essential oils, etc.).
6. 96-well microplate.
7. Vortex mixer.
8. Incubator at 37°C with orbital shaker.
9. Microplate photometer at 595 nm.

2.3 Determination of the Minimum Bactericidal Concentration (MBC)

1. Bacterial cultures to be studied.
2. Culturing tools: micropipettes, pipette tips, plastic 1.5-mL tubes, multichannel micropipette, and L-shaped spreaders.

3. Laboratory weighing material: precision balance, spoon, and spatula.
4. Growth media: tryptic soya broth and agar supplemented with 0.6% yeast extract (or any other nutritive broth or agar).
5. Antimicrobial agents (food preservatives, biocides, essential oils, etc.).
6. 96-well microplate.
7. Vortex mixer.
8. Incubator at 37°C with orbital shaker.
9. Microplate photometer at 595 nm.

2.4 Growth Kinetics

1. Bacterial cultures to be studied.
2. Culturing tools: micropipettes, pipette tips, and L-shaped spreaders.
3. Laboratory weighing material: precision balance, spoon, and spatula.
4. Growth media: tryptic soya broth and agar supplemented with 0.6% yeast extract (or any other nutritive broth or agar).
5. Antimicrobial agents (antibiotics, food preservatives, biocides, essential oils, etc.).
6. Glass test tubes and caps.
7. Phosphate-buffered saline (PBS).
8. Vortex mixer.
9. Incubator.
10. Thermocouple connected to a data logger.

2.5 Survival Curves Against Antimicrobial Agents or Food Preservation Methods

1. Bacterial cultures to be studied.
2. Culturing tools: micropipettes, pipette tips, and L-shaped spreaders.
3. Laboratory weighing material: precision balance, spoon, and spatula.
4. Growth media: tryptic soya agar supplemented with 0.6% yeast extract (or any other nutritive broth or agar).
5. Antimicrobial agents (antibiotics, food preservatives, biocides, essential oils, etc.) and equipment for food preservation treatments (i.e., thermoblock, pulse electric field generator, or high-pressure rig).
6. Plastic 1.5-mL tubes.
7. Sterile glass flask narrow neck and screw caps.

8. Washing/treatment media (sterile distilled water, phosphate-buffered saline, etc.).
9. Microcentrifuge.
10. Vortex mixer.
11. Incubator at 37°C.

3 Methods

3.1 Preparation of the Bacterial Working Culture

Before either of the methodologies, obtaining a working bacterial culture of the parental strain and a bacterial culture of each study strain is necessary. The working bacterial culture is obtained by following the steps below (see Fig. 1).

1. From a cryovial of the corresponding strain, inoculate and streak over the agar plate surface with an inoculation loop to isolate individual colonies.
2. Incubate the agar plates for 24 h at 37°C in aerobic conditions and in an inverted position (*see* **Note 1**).
3. Inoculate a single colony in 5 mL of growth broth in a test tube (*see* **Note 2**).
4. Incubate overnight at 37°C and 160 rpm.
5. Inoculate 10 μL from the overnight culture into 10 mL of growth broth in a flask.
6. Incubate at 37°C and 160 rpm until a stationary phase culture is obtained (*see* **Note 3**).

All experiments should be conducted with both the parental strain and study strains from ALEs or genetic engineering to compare results and to determine increases in tolerance and resistance. In addition, results should come from at least three replicates carried out with different bacterial cultures and preferably on different days (*see* **Note 4**).

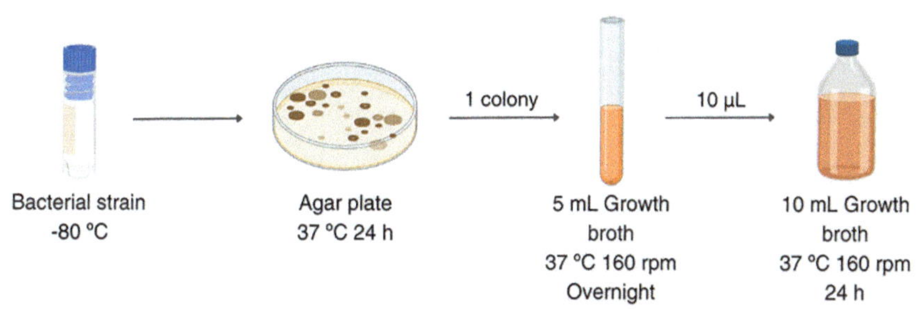

Fig. 1 Diagram for obtaining working bacterial culture

3.2 Determination of the Minimum Inhibitory Concentration (MIC)

This protocol aims to determine the minimum inhibitory concentration (MIC), the lowest concentration of an antimicrobial agent that inhibits bacterial growth.

This methodology is based on CLSI guidelines [24] for MIC determination by broth microdilution method. Furthermore, the MIC is assessed to reduce 90% of the antimicrobial growth measured by optical density (OD_{595}) compared to the positive control.

To carry out the MIC determination, the following steps should be taken:

1. Prepare a stock of the antimicrobial agent that is 10-fold more concentrated than the higher concentration to be tested (*see* **Note 5**).
2. In a microplate, dispense 100 μL of growth media into each well. Add an additional 80 μL more in the first well.
3. Homogenize the antimicrobial agent stock in a vortex and add 20 μL to the first well. Mix with a pipette tip.
4. From the first well perform serial dilutions by passing 100 μL from one well to the next homogenizing with the pipette tip. Repeat this process until all dilutions have been made (*see* **Note 6**) (see Fig. 2).
5. Homogenize with a pipette tip and remove 10 μL from each well.
6. Add 10 μL of bacterial culture (prepared in accordance with section 3.1.), previously diluted to achieve a concentration of approximately 3×10^5 CFU/mL (colony-forming units/mL) in the microplate well. Homogenize with pipette tip.
7. For controls dispense 90 μL of growth media in the microplate and 10 μL of the antimicrobial agent stock for negative control

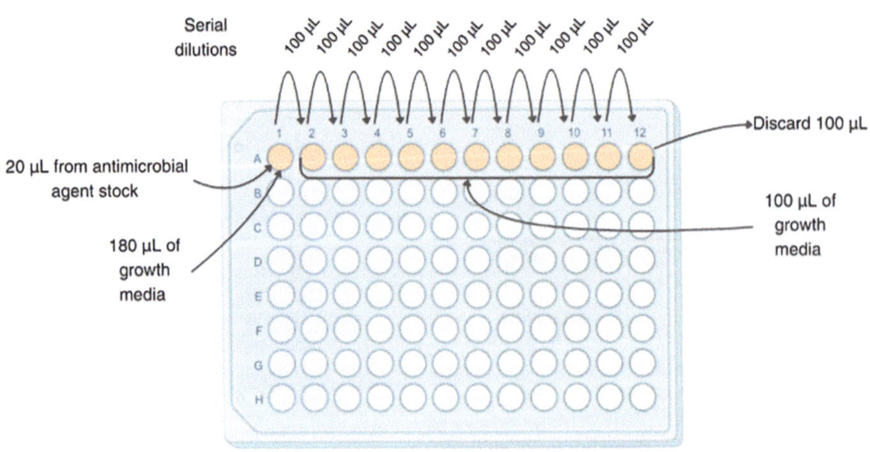

Fig. 2 Diagram of the preparation of the range of concentrations in microplate

and 10 μL of bacterial culture, previously diluted, for positive control. Homogenize with pipette tip.

8. Incubate the microplate for 24 h at 37°C at 160 rpm in aerobic conditions (*see* **Notes 1 and** 7).
9. Read optical density at a 595-nm wavelength (OD_{595}) in a microplate photometer, shaking it beforehand (*see* **Note 8**).
10. In the data table, subtract the negative control value to eliminate the effect of the culture media and the antimicrobial agent on the OD_{595} measurement.
11. Compare the OD_{595} value of each concentration with 10% of the OD_{595} value of the positive control [19]. The lowest concentration that achieves a 90% reduction in OD_{595} compared to the positive control will be the MIC.
12. Compare the MIC of parental strain with the MIC of study strains to detect RVs with increased resistance (see Fig. 3).

An increase in MIC suggests that the study strain has adapted to the antimicrobial agent, resulting in increased resistance and enhanced ability to grow in the presence of the antimicrobial agent.

3.3 Determination of the Minimum Bactericidal Concentration (MBC)

This protocol aims to determine the minimum bactericidal concentration (MBC), the lowest concentration of an antimicrobial agent that inactivates 99.9% of a bacterial population of approximately 3×10^5 CFU/mL.

To carry out the MBC determination, the following steps should be taken (continue from Step 8 of section 3.2.):

1. From each well of the microplate where growth has been inhibited, two different dilutions are seeded on an agar plate to asses inhibition (*see* **Note 9**):
   a. Plate 50 μL directly from the well, previously homogenized with the pipette tip, on the surface of an agar and spread with an L-shaped spreader.
   b. Homogenize with the pipette tip and take 10 μL for dilution in a plastic 1.5-mL tube with 990 μL of PBS.

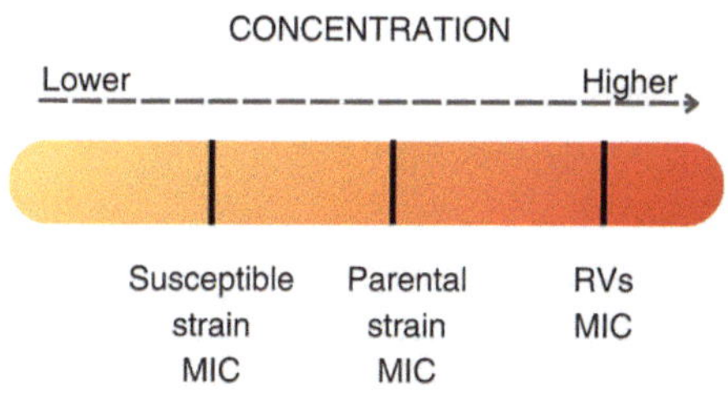

Fig. 3 Results interpretation of a minimum inhibitory concentration (MIC) assay

Then, seeds 100 μL on the surface of an agar plate from the 1.5-mL tube and spread with an L-shaped spreader.

2. Incubate agar plates for 24 h at 37°C in aerobic conditions and in an inverted position (*see* **Note 1**) (see Fig. 4).
3. Count the number of colonies on the agar plates (*see* **Notes 2 and 10**).
4. Calculate the CFU/mL and determine inactivated logarithmic cycles compared to initial bacterial population of approximately 3×10^5 CFU/mL (*see* **Note 11**).
5. The lowest concentration of antimicrobial agent that achieves at least 99.9% bacterial population reduction (3 $\log_{10}$ cycles) is the MBC.
6. Compare parental strain MBC with MBC of study strains to detect RVs with increased tolerance.

An increase in MBC suggests that the study strain has adapted to the antimicrobial agent, increasing its tolerance and capacity to survive in the presence of the antimicrobial agent.

3.4 Growth Kinetics

This protocol aims to determine the improvement or worsening in the growth fitness of study strains compared to the parental strain in the presence of the antimicrobial agent. For this purpose, the growth curve can be analyzed at different temperatures, such as optimum growth (e.g., 37°C) or refrigeration (i.e., 8°C). However, in both conditions, the experiment begins by following the steps below (see Fig. 5):

1. Prepare test tubes with 5 mL of growth media containing a subinhibitory concentration of antimicrobial agent, generally $\frac{1}{2} \times MIC$ (*see* **Note 12**). In addition, a positive control (no antimicrobial) and negative control (no microorganism) shall be placed.
2. Add 50 μL of bacterial culture (prepared in accordance with section 3.1.), previously diluted to obtain a final concentration of approximately 3×10^3 CFU/mL in the test tube. Homogenize using a vortex.

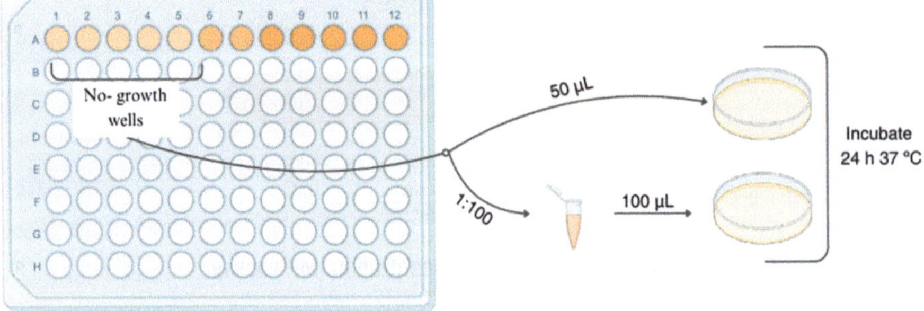

Fig. 4 Diagram of plate seeding to determine minimum bactericide concentration (MBC)

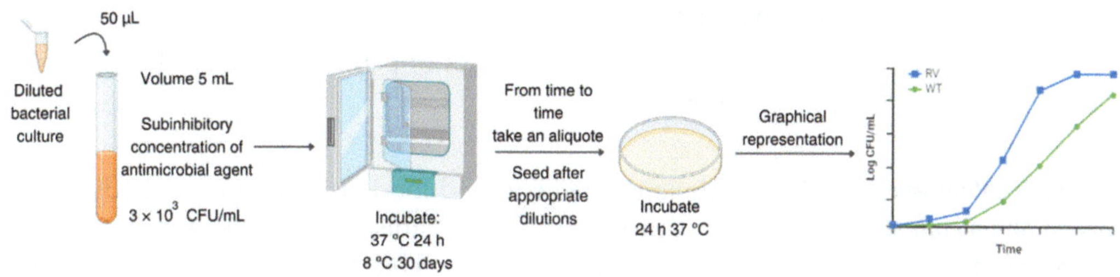

Fig. 5 Diagram of the elaboration of a growth kinetic

3. Place the thermocouple in the incubator to monitor the temperature.
4. Incubate at 37°C for 24 h (optimum growth) or at 8°C for 30 days (refrigeration) (*see* **Note 13**).
5. Take aliquots at specific treatment times and make the appropriate dilutions for plate seeding. Less dilutions are needed in the initial phases of growth, but it is necessary to adjust them according to the growth (*see* **Notes 14 and 15**).
6. At each specific treatment time selected, dilute and homogenize in a vortex and seeds 100 µL on the surface of an agar plate and spread with an L-shaped spreader. It is always necessary to make time 0.
7. Incubate agar plates at 37°C for 24 h in aerobic condition and in inverted position (*see* **Note 1**).
8. Count the number of colonies on the agar plates, and calculate the the CFU/mL (*see* **Notes 2, 10 and 11**).
9. Display graphically the bacterial growth curves through the representation of the $\log_{10}$ CFU/mL versus incubation time.
10. Model the survival curve using the following modified Gompertz equation [25]:

$$y = y_0 + A\exp\left\{-\exp\left[\left(\mu m\frac{e}{A}\right)(\lambda - t) + 1\right]\right\}$$

where *y*: $\log_{10}$ CFU/mL; y_0: initial population ($\log_{10}$ CFU/mL); *A*: difference between maximum value reached and initial population ($\log_{10}$ CFU/mL); μ_m: maximum growth rate (1/h); λ: lag time/time to detection (h); *t*: time (h) (*see* **Note 16**).

11. Apply at least squares adjustment using a statistical analysis and data visualization software. Evaluate the adjustment's goodness of fit by applying standard error (R^2), R^2 adjusted values, and *RMSE* (the root mean square error).

12. Obtain μ_m and λ values.
13. Compare μ_m and λ values of parental strain to those of study strains.

3.5 Survival Curves Against Antimicrobial Agents or Food Preservation Methods

Many antimicrobial agents and food preservation methods seek to inactivate pathogenic microorganisms through short-term treatments under lethal conditions. This protocol aims to analyze and evaluate the survival of a microbial population against the action of an antimicrobial compound by plotting microbial inactivation, on a logarithmic scale, over time of exposure to the antimicrobial agent applied at a constant intensity, i.e., constant dose of an antimicrobial, constant temperature in a heat treatment, etc.

To carry out a survival curve experiment, the following steps should be taken (see Fig. 6):

1. Prepare 1.5-mL plastic tubes with 1 mL of bacterial culture (prepared in accordance with section 3.1. *Preparation of the Bacterial Working Culture*) at the initial bacterial concentration (N_0) established for the treatment (approximately 3×10^9 or 3×10^7 CFU/mL).
2. Centrifuge 1.5-mL plastic tubes in the microcentrifuge at 12,000 rpm for 2 min. Carefully remove the supernatant while leaving the pellet at the bottom of the tubes.
3. Resuspend the pellet in 1 mL of washing media (*see* **Note 17**).
4. Repeat Step 2 by centrifuging and removing the supernatant.
5. Resuspend the pellet in 1 mL of treatment media.

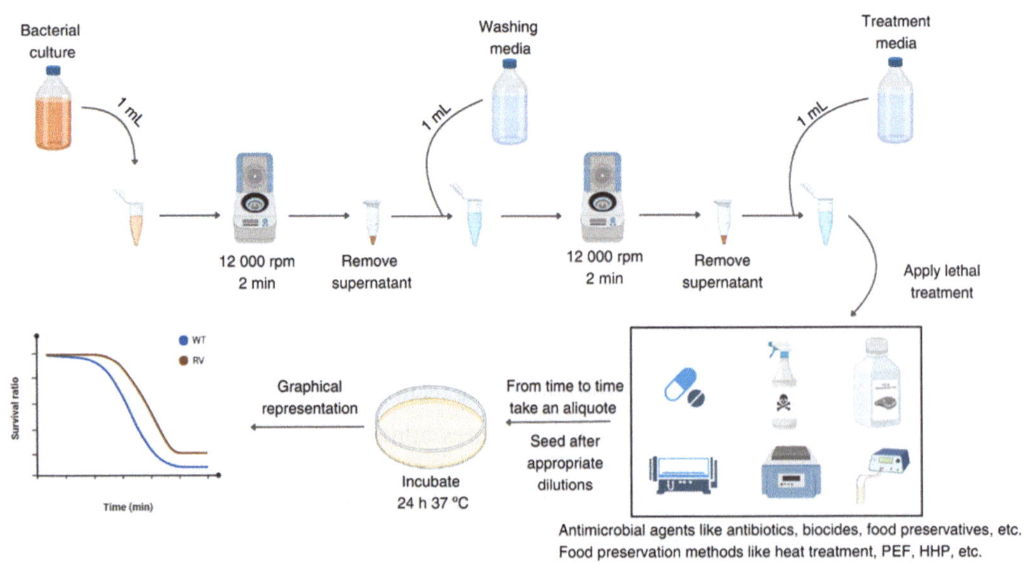

Fig. 6 Diagram of a lethal treatment with an antimicrobial agent or food preservation treatment

6. Apply a lethal treatment with an antimicrobial agent or food preservation method (*see* **Note 18**).
7. Take aliquots at specific treatment times and make the appropriate dilutions for plate seeding. More dilutions are needed in the initial phases of treatment, but it is necessary to adjust them according to the treatment.
8. At each specific treatment time selected, dilute and homogenize in a vortex and seeds 100 μL on the surface of an agar plate and spread with an L-shaped spreader (*see* **Note 19**). It is always necessary to make time 0.
9. Incubate agar plates at 37°C for 24 h in aerobic condition and in inverted position (*see* **Note 1**).
10. Count the number of colonies on the agar plates, and calculate the CFU/mL at each stage of treatment (N_t) (*see* **Notes 2, 10, and 11**).
11. Display graphically the bacterial survival curves through the representation of the survival ratio $\left(\log_{10} \frac{N_t}{N_0}\right)$ versus treatment time.
12. Model the survival curve using, for instance, the following Weibull equation [26]:

$$\log_{10} S(t) = \left(\frac{t}{\delta}\right)^p$$

where $S(t)$: survival ratio $\left(\frac{N}{N_0}\right)$; t: treatment time; δ: reduction time of the first logarithmic cycle; p: shape factor (>1 convex; <1 concave; =1 linear) (*see* **Note 20**).
13. Apply at least squares adjustment using a statistical analysis and data visualization software. Evaluate the adjustment's goodness of fit by applying standard error (R^2), R^2 adjusted values, and *RMSE* (the root mean square error).
14. Obtain δ and p values.
15. Calculate the time required to inactivate 3 cycles (D_3) of both the parental strain and the study strains, in order to detect RVs with increased tolerance to lethal treatments.

Although these methodologies facilitate the understanding of bacteria protection mechanisms, it also allow us to evaluate the ability of bacteria to develop tolerance/resistance against antimicrobial agents and food preservation method. Genotypic characterization is also needed to identify the genetic determinants responsible for increased protection and to study the interactions between the different mutated genes, which may lead to synergistic effects that explain the increase in tolerance/resistance or, in the opposite case, compensatory effects that limits the development of RVs.

These experiments are carried out in an in vitro environment and further study is needed to determine the potential impact on food safety.

4 Notes

1. Incubation temperature, time, and oxygen availability should be modified according to the optimal growth conditions of the bacteria.
2. Check the size and shape of the colonies, and verify the homogeneity in the agar plate to avoid microbial contamination.
3. To obtain a stationary phase culture, incubation temperature and time should be modified according to the optimal growth conditions of the bacteria.
4. At least 3 replicates are recommended to ensure the reproducibility of the experiment and the reliability of the results.
5. Some antibiotics are photosensitive, so the test must be carried out in low light.
6. To evaluate a tighter range of concentrations than the CLSI methodology (typically, ranges of 10%, 25%, or 50% are assessed), it is necessary to prepare a stock 10-fold more concentrated for each of the concentrations to be evaluated. Methodologically replace Steps 2–4 of Determination of the Minimum Inhibitory Concentration (MIC) with:
   a. In the microplate dispense 90 μL of growth media.
   b. Homogenize the antimicrobial agent stock in a vortex and add 10 μL to the corresponding well. Homogenize with pipette tip.
   c. Repeat step b for each tested concentration.
7. The contents of the wells of a microplate can evaporate, especially in incubations >24 h, so it is recommended to cover the plate with a sterile film with a hole in each well to reduce evaporation.
8. For the determination of the optical density, a wavelength between 590 and 620 nm can be used.
9. Food preservatives generally have no bactericidal capacity as they have the function of inhibiting bacterial growth but do not inactivate.
10. For an adequate plate surface count there must be between 10 and 300 colonies. The upper limit may be increased if equipment such as automatic plate counters is used after validation.

11. To calculate the CFU/mL, multiply the microbial count by the dilution factor and the seeding factor.
12. In products whose use is regulated, such as food preservatives, the legal limits may be used if they are sub-inhibitory doses. References doses for use in products such as antibiotics or even disinfectants can also be used if they exist.
13. Times and temperatures in the growth kinetics must be adjusted for each microorganism and antimicrobial used.
14. The time taken to take the aliquot should be as short as possible to avoid affecting the temperature of the culture during the experiment.
15. The time between aliquots does not have to be constant and should be adjusted to the growth of the bacteria. In treatments at optimal growth conditions of the bacteria it is more interesting to monitor the first hours of growth, to correctly determine the lag phase (λ) and the maximum growth rate in the exponential phase (μ_m); however, in treatments at cooling temperatures, aliquots can be taken every few days and increase the frequency when growth begins to be observed.
16. The Gompertz model assumes that conditions (temperature, pH, nutrient availability, etc.) are constant.
17. This step also has the function of removing organic matter from the growth media.
18. Typically, a treatment time of 10–30 min is desired for applicability and repeatability. In addition, some antimicrobial agents need specific conditions:
    - Biocide efficacy depends on the treatment temperature, so both the products and the treatment must be at a controlled temperature.
19. In lethal treatments with some antimicrobial agents, like biocides or essential oils, it is necessary to stop the residual action of the antimicrobial agent after the extraction of the alicuot at the preset interval for which the alicuot must be neutralized. Specific neutralizers are available for some products, such as sodium thiosulphate for chlorine-based biocides, but generic neutralizers such as buffered peptone water or phosphate-buffered saline can be used.
20. The Weibull model is not able to accurately describe sigmoid survival curves (which transition from convex to concave or vice versa), but it is effective in modeling most observed survival curves, including those that are concave, linear, or convex.

References

1. European Food Safety Authority, European Centre for Disease Prevention and Control (2024) The European Union One Health 2023 Zoonoses report. EFSA J 22:e9106
2. Munita JM, Arias CA (2016) Mechanisms of antibiotic resistance. Microbiol Spectr J 4:37
3. Brauner A, Fridman O, Gefen O, Balaban NQ (2016) Distinguishing between resistance, tolerance and persistence to antibiotic treatment. Nat Rev Microbiol 14:320–330
4. Capita R, Alonso Calleja C (2013) Antibiotic-resistant bacteria: a challenge for the food industry. Crit Rev Food Sci Nutr 53:11–48
5. Nicoloff H, Perreten V, Levy SB (2007) Increased genome instability in *Escherichia coli* lon mutants: relation to emergence of multiple-antibiotic-resistant (Mar) mutants caused by insertion sequence elements and large tandem genomic amplifications. Antimicrob Agents Chemother 51:1293–1303
6. Karatzas KAG, Randall LP, Webber M, Piddock LJV, Humphrey TJ, Woodward MJ, Coldham NG (2008) Phenotypic and proteomic characterization of multiply antibiotic-resistant variants of *Salmonella enterica* serovar Typhimurium selected following exposure to disinfectants. Appl Environ Microbiol 74:1508–1516
7. Rizzotti L, Rossi F, Torriani S (2016) Biocide and antibiotic resistance of *Enterococcus faecalis* and *Enterococcus faecium* isolated from the swine meat chain. Food Microbiol 60:160–164
8. Paul D, Chakraborty R, Mandal SM (2019) Biocides and health-care agents are more than just antibiotics: inducing cross to co-resistance in microbes. Ecotoxicol Environ Saf 174:601–610
9. Roedel A, Dieckmann R, Brendebach H, Hammerl JA, Kleta S, Noll M, Dahouk SA, Vincze S (2019) Biocide-tolerant *Listeria monocytogenes* isolates from german food production plants do not show cross-resistance to clinically relevant antibiotics. Appl Environ Microbiol 85:20
10. Chueca B, Berdejo D, Gomes-Neto NJ, Pagán R, García-Gonzalo D (2016) Emergence of hyper-resistant *Escherichia coli* MG1655 derivative strains after applying sub-inhibitory doses of individual constituents of essential oils. Front Microbiol 7:273
11. Berdejo D, Merino N, Pagán E, García-Gonzalo D, Pagán R (2020) Genetic variants and phenotypic characteristics of *Salmonella* Typhimurium-resistant mutants after exposure to carvacrol. Microorganisms 8:22
12. Giacometti F, Shirzad-Aski H, Ferreira S (2021) Antimicrobials and food-related stresses as selective factors for antibiotic resistance along the farm to fork continuum. Antibiotics 10:29
13. Nasr AM, Mostafa MS, Arnaout HH, Elshimy AAA (2018) The effect of exposure to sub-inhibitory concentrations of hypochlorite and quaternary ammonium compounds on antimicrobial susceptibility of *Pseudomonas aeruginosa*. Am J Infect Control 46:57–63
14. Chueca B, Renzoni A, Berdejo D, Pagán R, Kelley WL, García-Gonzalo D (2018) Whole-Genome sequencing and genetic analysis reveal novel stress responses to individual constituents of essential oils in *Escherichia coli*. Appl Environ Microbiol 84:7
15. Ma Y, Lan G, Li C, Cambaza EM, Liu D, Ye X et al. (2019) Stress tolerance of *Staphylococcus aureus* with different antibiotic resistance profiles. Microb Pathog 103549:6
16. Uddin MJ, Jeon G, Ahn J (2018) Variability in the adaptive response of antibiotic-resistant *Salmonella* Typhimurium to environmental stresses. Microb Drug Resist 25:182–192
17. Pagán E, Pavli F, Happiette S, Berdejo D, Gatt R, Pagán R, Valdramidis V, García-Gonzalo D (2024) Isolation and characterization of resistant variants of *Salmonella* Typhimurium after sequential exposure to plasma activated water (PAW). Innov Food Sci Emerg Technol 93:103633
18. Oniciuc EA, Likotrafiti E, Alvarez-Molina A, Prieto M, López M, Alvarez-Ordóñez A (2019) Food processing as a risk factor for antimicrobial resistance spread along the food chain. Curr Opin Food Sci 30:21–26
19. Kohanski MA, DePristo MA, Collins JJ (2010) Sublethal antibiotic treatment leads to multidrug resistance via radical-induced mutagenesis. Mol Cell 37:311–320
20. Berdejo D, Pagán E, Merino N, García-Gonzalo D, Pagán R (2021) Evolution assays for the isolation of mutant bacteria resistant to natural antimicrobials. In: Magnani M (Ed.)Eds. Detection and enumeration of bacteria, yeast, viruses, and protozoan in foods and freshwater. Springer US, New York, NY, 65–75
21. Wong A, Rodrigue N, Kassen R (2012) Genomics of adaptation during experimental evolution of the opportunistic pathogen *Pseudomonas aeruginosa*. PLOS Genet 8:13
22. O'Callaghan D, Charbit A (1990) High efficiency transformation of *Salmonella*

Typhimurium and *Salmonella* typhi by electroporation. Mol Gen Genet 223:156–158
23. Li XT, Thomason LC, Sawitzke JA, Costantino N, Court DL (2013) Positive and negative selection using the tetA-sacB cassette: recombineering and P1 transduction in *Escherichia coli*. Nucleic Acids Res 41:e204
24. Clinical and Laboratory Standards Institute (2015) Methods for dilution antimicrobial susceptibility tests for bacteria that grow aerobically, 10th ed. CLSI DOCUMENT M07-A10. Wayne, Pennsylvania, USA: Clinical and Laboratory Standards Institute
25. Zwietering MH, Jongenburger I, Rombouts FM, van 't Riet K (1990) Modeling of the bacterial growth curve. Appl Environ Microbiol 56:1875–1881
26. Mafart P, Couvert O, Gaillard S, Leguerinel I (2002) On calculating sterility in thermal preservation methods: application of the Weibull frequency distribution model. Int J Food Microbiol 72:107–113

Chapter 8

Detection of the Protozoan *Toxoplasma gondii* in Meat and Meat Products

Daniel Berdejo, Regina Lázaro, Nabil Halaihel and Susana Bayarri

Abstract

In this chapter, we describe a methodology for detecting viable *Toxoplasma gondii* in meat and meat products. The method compromises two main steps: (1) the detection and quantification of *T. gondii* DNA using the quantitative polymerase chain reaction technique (qPCR); and (2) the evaluation of parasite viability through a mouse bioassay, in which serological analysis using immunofluorescence assay (IFA) and qPCR analysis of brain tissue are performed. To enhance the sensitivity of the method, a concentration technique based on acid pepsin digestion is applied prior to both protocols (qPCR and bioassay analyses).

Keywords *Toxoplasma gondii*, Meat and meat products, Pepsin digestion, qPCR, Mouse bioassay, IFA

1 Introduction

The obligate intracellular protozoan *Toxoplasma gondii* is the causative agent of toxoplasmosis, one of the most frequent parasitic infections in humans and other warm-blooded animals. It has been found worldwide and approximately one-third of global population has been exposed to this parasite [1, 2]. Infection in humans is mostly asymptomatic, but it can cause blindness and intellectual disability in congenitally infected children and devastating disease in immunocompromised individuals [3].

The European Food Safety Authority (EFSA) identifies *T. gondii* as one of the most relevant parasites among all pathogens listed as the main agents of food-borne diseases, and estimates that food-borne transmission accounts for 40–60% of all *T. gondii* infections [4]. Specifically, the consumption of undercooked or raw meat from livestock that contains viable cysts (bradyzoites) has been identified as a primary source of infection [5–9]. Additionally, EFSA highlights *Toxoplasma* as one of the most significant biological hazards in meat inspection. However, current regulations do not establish criteria

Marciane Magnani and Geany Targino de Souza Pedrosa (eds.), *Detection and Quantification of Bacteria, Yeast, Viruses, and Protozoan in Foods and Freshwater*, Methods and Protocols in Food Science, https://doi.org/10.1007/978-1-0716-5064-6_8,

for its detection during official meat inspection at slaughterhouses [10]. For this reason, the determination of viable *T. gondii* in meat and meat products is a necessity for both food operators in the meat industry and competent authorities to assess and manage the risk to human health.

Nevertheless, detection of viable *T. gondii* in meat and meat products is challenging and often requires highly sensitive methods due to the parasite's typically low concentration, around one cyst per 100 g of meat [11], and heterogeneous distribution in animal tissues. Therefore, false negatives can result from insufficient sample size or improper sampling [12, 13].

One of the most accurate current techniques for the detection of *T. gondii* is quantitative polymerase chain reaction (qPCR). The qPCR is a molecular biology technique used to amplify and simultaneously quantify a specific DNA sequence. It allows the measurement of target DNA in real time by using fluorescent dyes or probes, providing both qualitative and quantitative data. The most common targets for identifying *T. gondii* through qPCR are the B1 fragment (which appears 35 times in the parasite's genome) and a 529 bp region (that is repeated 200–300 times). However, since the standard commercial DNA isolation methods are optimized for 25 mg of sample, the likelihood of detecting the parasite in meat and meat products is low. A prior concentration stage, such as pepsin digestion, can also be included in order to increase sensitivity [14]. On the other hand, while the presence of *T. gondii* DNA in the sample extracts indicates that the meat comes from an infected animal, it does not necessarily imply that the product contains infectious organisms [15].

Bioassay is considered the reference method for detecting infective *T. gondii* in meat or meat products. It involves the use of laboratory animals, such as mice. Meat samples are digested with acid pepsin or trypsin and a fraction of the sediment is inoculated in mice, generally intraperitoneally or subcutaneously [16–18]. Typically, two to five mice are used per sample, maintained in observation for 6–8 weeks, and they are tested for antibodies to *T. gondii*. If the parasite is viable in the meat sample, tissue cysts can form in the brain of exposed mice (Fig. 1) [19]. This can be confirmed by microscopy or qPCR [12, 20].

Although bioassay is a sensitive technique to determine the infectivity of *T. gondii*, it requires the use of live animals, which is not desirable for screening large numbers of samples from an animal ethics point of view. Moreover, it is time-consuming and resource-intensive, making it unsuitable for routine testing [7, 13, 21]. For this reason, molecular techniques are used for routine detection of the presence of *T. gondii* DNA in samples, whereas bioassays are applied to positive samples to obtain information on the viability of *Toxoplasma* [15].

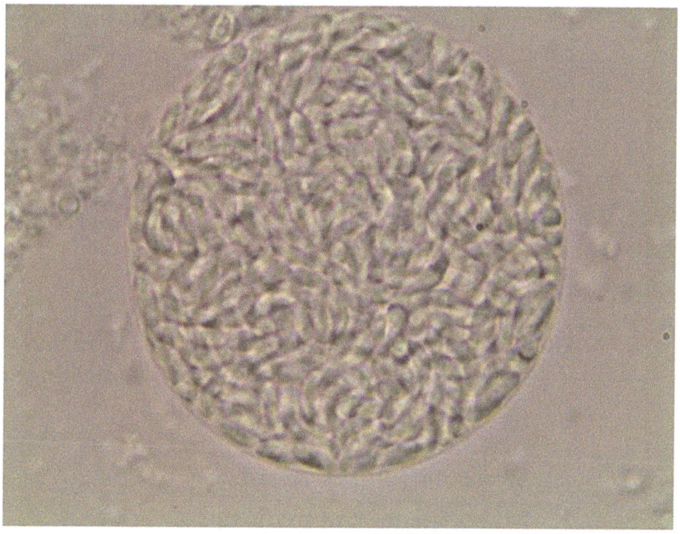

Fig. 1 *Toxoplasma gondii* cyst in mouse brain [19]

In this chapter we present a methodology for the detection of viable *Toxoplasma gondii* in meat and meat products. Firstly, samples are analyzed by real-time polymerase chain reaction technique (qPCR) for the detection and quantification of *T. gondii* DNA. Then, positive qPCR samples are analyzed by mouse bioassay to determine the viability of the parasite. Prior to both techniques, a concentration step based on an acid pepsin digestion is included.

2 Materials

2.1 Acid Pepsin Digestion

1. Analytical balance.
2. Knife, tongs, and spatula for sampling.
3. Beakers (1 L).
4. Vortex mixer.
5. Graduated cylinder, pipette filler, and pipettes.
6. Blender.
7. Erlenmeyer flask (1 L).
8. Saline solution: 0.85% (w/v) NaCl in distilled water.
9. Pepsin solution: 0.52% (w/v) pepsin (activity 1: 10,000), 1% (w/v) NaCl, 1.4% (v/v) HCl (25%) in distilled water.
10. pH meter.
11. Shaking water bath with temperature control.

12. Stainless steel mesh sieve with 180-μm pore size.
13. Centrifuge tubes.
14. Centrifuge.
15. PBS: phosphate-buffered saline at pH 7.2: 2.7 mM KCl, 140 mM NaCl, 10 mM phosphate.
16. Sodium bicarbonate solution at pH 8.3: 1.2% $NaHCO_3$ (w/v) in distilled water.
17. Distilled water.
18. Antibiotic saline solution: sterile 0.85% NaCl solution supplemented with 1000 U/mL penicillin and 100 mg/mL streptomycin.

2.2 qPCR Analysis

1. Sterile microtubes 1.5 mL.
2. Micropipettes and sterile tips.
3. Milli-Q sterile water.
4. Microcentrifuge.
5. 0.2 mL PCR tubes.
6. PCR reagents (TaqDNA polymerase, deoxynucleotides (dNTPs), and buffer solution).
7. Positive and negative controls for *T. gondii* and internal control, such as actin DNA copies.
8. Real-time PCR instrument and related equipment.
9. For protocol A: primers RocFw and RocRv (Table 1), and a fluorescent dye, such as SYBR green (if it is not included in master mix).

For protocol B: primers ToxoFw and ToxoRv, along with the probe ToxoProbe (Table 1).

2.3 Mouse Bioassay

1. CD1 Swiss mice or similar strains (25–40 g).
2. Animal house for experimental studies, laboratory animal facilities, mouse handling equipment.
3. Sterile microtubes 0.5 mL.
4. Syringes with needles (27 G and 23 G).
5. Antiseptic, such as 70% alcohol.
6. Micropipettes and sterile tips.
7. Microcentrifuge.
8. CO_2 chamber for mice euthanasia.
9. Sterile surgical supplies (forceps, scissors, tongs…).
10. Sterile microtubes 1.5 mL.

Table 1
Primers and probes used in qPCR for *T. gondii* detection and quantification

Protocol	Primers	Sequences (5′→ 3′)	Amplified region	Amplicon length (bp)
A	RocFw	TAGACGAGACGACGCTTTCC	Repeat element	529
	RocRv	TCGCCCTCTTCTCCACTCT		
B	ToxoFw	TCCCCTCTGCTGGCGAAAAGT	Glycerol-3-phosphate dehydrogenase (B1)	97
	ToxoRv	AGCGTTCGTGGTCAACTATCGATTG		
	ToxoProbe	FAM – CTGTGCAACTTTG GTGTATTCGCAG – BHQ		

2.4 Immunofluorescence Assay (IFA)

1. 0.2 mL microtubes.
2. Micropipettes and sterile tips.
3. Serum diluting buffer: phosphate buffer saline containing 1% (v/v) bovine serum albumin.
4. Slides kit for the detection of mouse-specific IgG antibodies against *Toxoplasma gondii*, labeled with fluorescein isothiocyanate (FITC).
5. Incubator chamber (37°C).
6. Phosphate-buffered saline at pH 9.0.
7. Phosphate-buffered saline at pH 7.2.
8. Anti-mouse IgG solution.
9. Coplin jar, magnetic stirrer, and magnetic stir bar.
10. Distilled water.
11. Slide mounting fluid.
12. 24 × 50 × 50 mm microscope cover slips.
13. Fluorescence microscope (UV-light) and related equipment.

3 Methods

3.1 Acid Pepsin Digestion of Meat and Meat Products

Pepsin digestion is first performed and involves treating the meat with a pepsin-HCl solution, which causes the rupture of the tissue cyst wall of *T. gondii* and the release of the bradyzoites inside. This step is crucial for concentrating the sample, thereby enabling the analysis of a larger amount of tissue (Fig. 2). Subsequently, qPCR is conducted to determine the presence of *T. gondii* DNA in the samples or mouse bioassay to confirm viability.

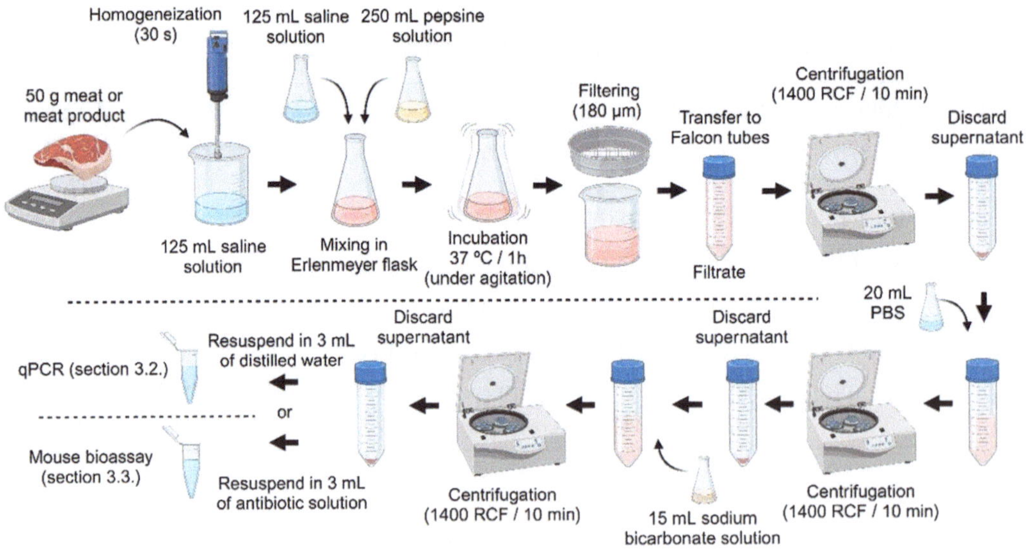

Fig. 2 Scheme of the acid pepsin digestion procedure

Protocol to carry out the acid pepsin digestion of meat and meat products:

1. Weigh 50 g of meat or meat product and place it in a beaker (*see* **Note 1**).
2. Add 125 mL of saline solution to the beaker (*see* **Note 2**).
3. Homogenize the mixture using a blender for approximately 30 s until a homogeneous consistency is achieved.
4. Rinse any remaining residue from the blender with 125 mL of additional saline solution and transfer the homogenized mixture into a 1 L Erlenmeyer flask.
5. Add 250 mL of pre-warmed pepsin solution (37°C) to the Erlenmeyer flask containing the tissue homogenized (*see* **Note 3**).
6. Incubate the Erlenmeyer flask in a shaking water bath at 37°C for 1 h with continuous agitation.
7. Filter the incubated mixture through a 180-µm-pore sieve and collect the filtrate in a beaker.
8. Transfer the filtrate to Falcon tubes and centrifuge at 1400 RCF for 10 min.
9. Discard the supernatant and resuspend the sediment in 20 mL of PBS.
10. Centrifuge the resuspended sediment at 1400 RCF for 10 min and discard the supernatant.

11. Add 15 mL of freshly prepared sodium bicarbonate solution to neutralize the pepsin (*see* **Note 4**).
12. Centrifuge the mixture at 1400 RCF for 10 min and discard the supernatant.
13. Resuspend the remaining sediment in 3 mL of sterile distilled water if intended for qPCR analysis, or in an antibiotic saline solution if intended for mouse bioassay, and then distribute the suspension into multiple aliquots in 1.5 mL tubes for easier handling (*see* **Note 5**).

3.2 qPCR Analysis to Detect and Quantify *T. gondii* DNA

To determine the presence and quantify the DNA of *T. gondii* in the sample, an aliquot of 200 μL from the digested samples obtained in Section 3.1 should be used for DNA extraction and purification using commercial kits, following the manufacturer's protocol. The extracted DNA can be preserved by freezing until qPCR is performed (Fig. 3).

For the detection and quantification of *T. gondii* DNA, two distinct qPCR assays and set of primers can be used (Table 1), each targeting a different sequence: A) one assay uses a pair of primers with dyes as probes for detecting the specific 529 bp repetitive element, and B) the other uses a TaqMan probe for the glycerol-3-phosphate dehydrogenase (B1) gene (*see* **Note 6**).

Once the DNA has been extracted and purified, proceed with qPCR according to the following protocol:

1. Prepare PCR tubes for each sample, and for positive and negative controls. Prepare also a separate reaction for actin DNA copies as internal control (IC) (*see* **Note 7**).
2. Add all the reagents listed in Table 2, following either Protocol A or Protocol B, into the PCR tubes in the specified order.
3. Seal the plate and place it in thermal cycler.
4. Set up the thermal cycler with the parameters specified in Table 3, following either protocol A or protocol B.
5. Once the PCR has been completed, obtain the Ct values for each sample (or the average for each measurement).
6. Ensure that a signal is observed at 81°C on the dissociation curve to confirm the specificity of the amplification of the target repetitive element (protocol A) (*see* **Note 8**).

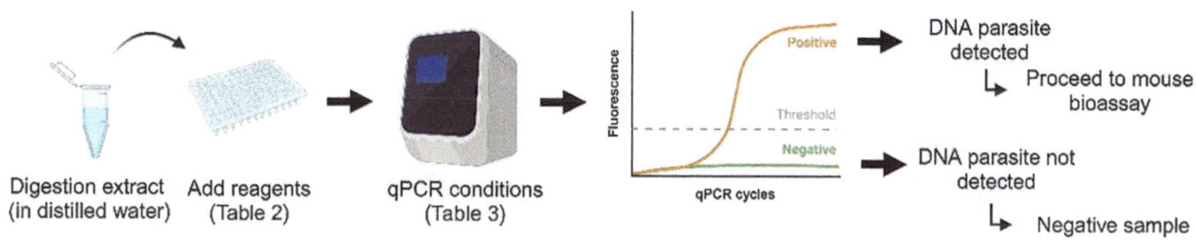

Fig. 3 Scheme of qPCR analysis to detect *T. gondii* DNA

Table 2
Reagents and concentrations used for qPCR according to protocols A and B

Protocol	Reagent	Volume (μL) for 20 μL reaction	Final concentration
A	qPCR Master Mix (2×)	10	1×
	RocFw (10 μM)	1	0.5 μM
	RocRv (10 μM)	1	0.5 μM
	Template DNA (sample)	4	10-100 ng/20 μL
	SYBR green	0.5	1 μM
	Hot-Start Taq DNA polymerase (5 U/μL)	0.2	1 unit/20 μL
	Nuclease-free water	Added up to a total volume of 20 μL	
B	qPCR Master Mix (2×)	10	1×
	Toxo Fw (10 μM)	1	0.5 μM
	Toxo Rv (10 μM)	1	0.5 μM
	ToxoProbe (6 μM)	1	0.3 μM
	Template DNA (sample)	4	10–100 ng/20 μL
	Hot-Start Taq DNA polymerase (5 U/μL)	0.2	1 unit/20 μL
	Nuclease-free water	Added up to a total volume of 20 μL	

Table 3
Parameters of the qPCR steps according to protocols A and B

Protocol	Cycling step	Temperature (°C)	Holding time	Number of cycles
A	Enzyme activation	94	7 min	1
	Denaturation	94	5 s	44
	Annealing	55	30 s	
	Extension	72	10 s	
	Final extension	72	10 min	1
	Dissociation curve	60 to 94 (Δ 0.5)	–	1
B	Enzyme activation	94	5 min	1
	Denaturation	94	5 s	44
	Annealing	55	30 s	
	End	10	–	1

Based on the protocols described and the specific conditions of our laboratory, the threshold cycle (Ct) for sample positivity is set at 38 for protocol A and at 32 for protocol B. However, it is crucial to establish the specific limit of detection (LOD) for each laboratory during the optimization of the technique.

By preparing calibration curves between Ct values and DNA copies under the specific conditions of the laboratory, it is possible to quantify the parasitic load (*see* **Note 9**). Once the Ct values of the sample have been interpolated to determine the number of DNA copies, the parasitic load can also be calculated. To achieve this, it is necessary to consider that the repetitive element (protocol A) is present 200–300 times per cell, while the B1 gene fragment appears 35 times in the parasite's genome. Finally, the weight of the sample for pepsin digestion, as well as the amount of sample used for qPCR, must be considered to adjust the dilutions and accurately calculate the parasitic load in the sample.

qPCR-positive samples indicate the presence of DNA. However, the parasite may not be viable and, therefore, not infectious. For this reason, a viability analysis should be performed using a mouse bioassay with qPCR-positive samples.

3.3 Mouse Bioassay to Evaluate Viability of *T. gondii*

To evaluate the viability of *T. gondii*, a mouse bioassay is conducted (Fig. 4). The following procedure requires approval from the relevant ethics committees and authorization from the appropriate institutions to conduct animal testing. Additionally, personnel must hold the appropriate accreditation to handle mice, conduct experiments, and perform euthanasia.

Typically, two to five mice are used per sample, which are inoculated intraperitoneally and observed for a period of 8 weeks. If meat contains *Toxoplasma* cysts, inoculated mice may develop an immune response detectable through serological analysis. If the parasite is viable, tissue cysts may form in the brain of exposed mice. To confirm this, the mice are tested for antibodies to *T. gondii* using the immunofluorescence assay technique (IFA). After euthanasia, brain is analyzed for the presence of *T. gondii* using qPCR. During this period, the mice must be cared for and kept under appropriate conditions in accredited facilities, in accordance with current legal and ethical regulations. Negative control mice are included in all experiments to ensure the validity of the results.

The following procedure is used to conduct the mouse bioassay:

(a) **Intraperitoneal inoculation.**

1. Load a syringe with 0.5 mL of the digestion extract obtained in Step 13 of Section 3.1, dissolved in antibiotic saline solution, and attach a needle (approximately 25–27 G).

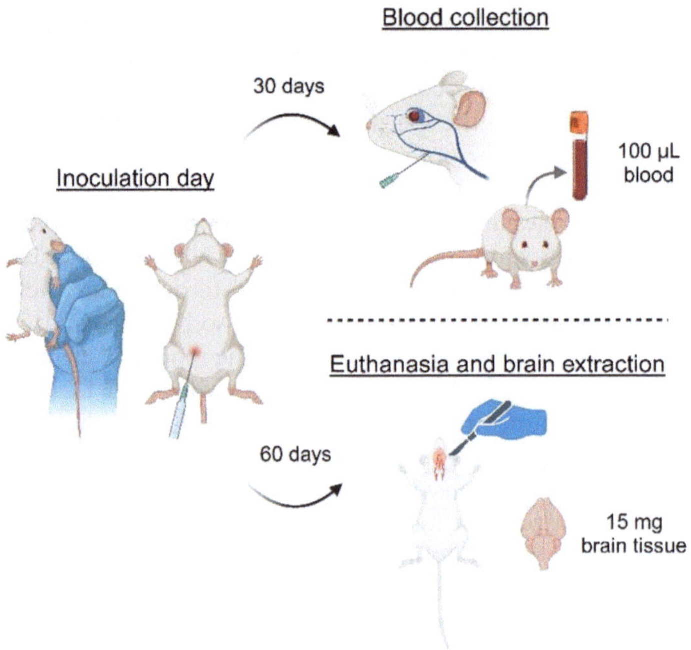

Fig. 4 Scheme of mouse bioassay

2. Secure the mouse and disinfect the peritoneal region at the site of injection.
3. Allow the area to dry for a few seconds.
4. Hold the syringe with the needle bevel facing upward at a 15–20° angle relative to the abdomen and slightly tilt the mouse head-down to shift the abdominal organs upward, reducing the risk of puncturing them.
5. Insert the needle into the lower right or left quadrant, avoiding the midline to prevent damage to major blood vessels or organs, and penetrate approximately 2–3 mm deep, depending on the size of the mouse (*see* **Note 10**).
6. Slowly inject the solution to avoid causing pressure or discomfort (*see* **Note 5**).
7. Remove the needle gently and apply light pressure to the injection site with a sterile swab to prevent leakage.
8. Monitor the mouse for any signs of distress or adverse reactions, such as abnormal posture, labored breathing, or lethargy.
9. Return the mouse to its cage and ensure it is in a warm and stress-free environment and continue monitoring the mouse for at least 10–15 min to ensure no immediate complications arise.

(b) Blood collection (30 days after inoculation).

1. Hold the mouse securely by grasping the loose skin at the back of the neck and shoulders (scruffing).
2. Locate the submandibular vein (*see* **Note 11**).
3. Use a sterile needle to puncture the submandibular vein (approximately 21–23 G). Position the needle at a 45° angle to the cheek, aiming for the vein, and apply gentle pressure until the vein is punctured. Blood should flow immediately after the puncture.
4. Use a microtube to collect approximately 100 μL of blood, avoiding excessive handling to ensure the mouse remains calm.
5. Apply gentle pressure to the puncture site with a sterile cotton swab or gauze until the bleeding stops and inspect the area to ensure there are no complications, such as prolonged bleeding.
6. Return the mouse to its cage and monitor it for 10 min for signs of stress or complications, ensuring it is kept warm and comfortable.

(c) Euthanasia using a CO_2 chamber and brain extraction (60 days after inoculation).

1. Gently transfer the mouse into the CO_2 chamber to minimize stress, avoiding chasing or rough handling, and securely place the chamber lid to prevent gas leakage.
2. Gradually introduce CO_2 into the chamber at a rate of 30–70% of the chamber volume per minute, as recommended by ethical guidelines. Carefully observe the mouse, as unconsciousness typically occurs within 1–3 min. Continue CO_2 flow for at least 1 min after breathing has stopped to ensure death.
3. Remove the mouse from the chamber and confirm the absence of vital signs, such as breathing, heartbeat, or reflexes. Use physical methods, such as palpating the chest, checking the eyes, or observing for movement.
4. Once the mouse has been euthanized, disinfect the work area and prepare the surgical tools for brain extraction.
5. Place the mouse on its back on a clean, flat surface, or a dissection tray.
6. Using sterile scissors, make a small incision at the back of the neck and extend it along the midline of the head. Peel back the skin to expose the skull, securing it in place with forceps as needed.

7. Use sterile scissors or a scalpel to carefully cut through the cranium, starting at the base of the skull and moving toward the snout. Make shallow cuts to avoid damaging the brain tissue. Once the cranium has been opened, gently lift and remove the pieces of bone using forceps to fully expose the brain.
8. Using small, sterile spatula or forceps, gently lift the brain from the cranial cavity. Start at the base of the brainstem and work slowly to avoid tearing or damaging the tissue and place the brain in a microtube.
9. Dispose of the carcass and any contaminated materials according to institutional biohazard protocols.

3.3.1 Immunofluorescence Assay (IFA) of Mouse Sera

Serum samples of mice are analyzed by IFA to detect antibodies against *T. gondii* with polyclonal rabbit antimouse immunoglobulins (Fig. 5). In this way, it becomes possible to observe whether the mouse has developed an immune response to *T. gondii*. Serum samples with a titer of 1:10 are considered positive. Positive and negative controls should be included in all tests to ensure assay validity.

The following procedure is used to assess the seropositivity of the mice:

1. Keep the blood obtained in Step 4 (Sect. 3.3) undisturbed for 30–60 min to allow clot formation.
2. Centrifuge at approximately 2000 RCF for 10 min, and using a micropipette, carefully collect the serum supernatant using a micropipette and transfer it to a new microtube.

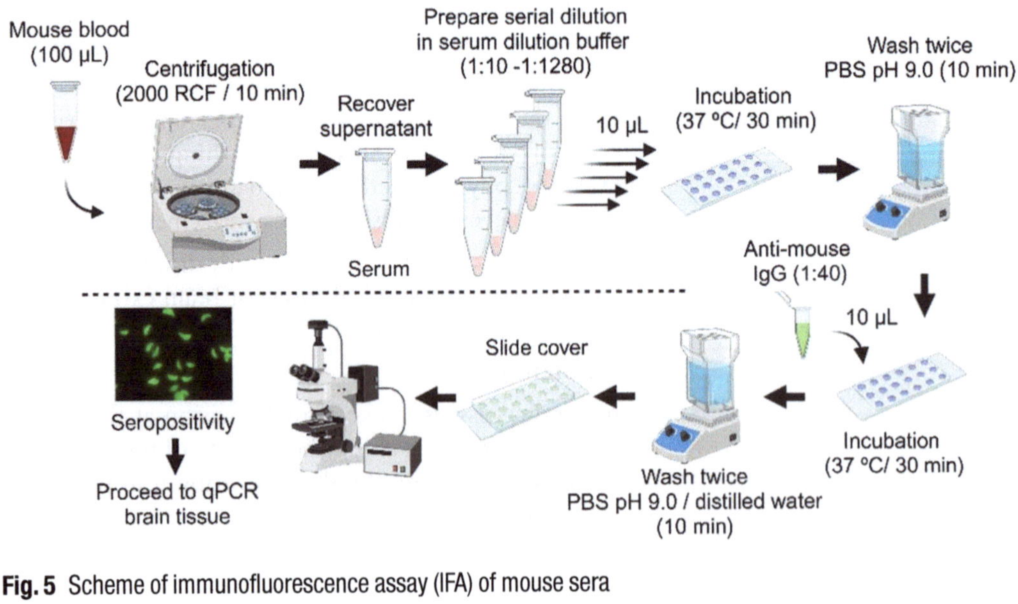

Fig. 5 Scheme of immunofluorescence assay (IFA) of mouse sera

3. Dilute the serum sample 1:10 using the serum diluting buffer.
4. Prepare serial dilution 1:2 from the initial dilution serum 1:10 up to 1:1280 using serum diluting buffer.
5. Add 10 μL of each serum dilution to the wells of *T. gondii* kit slides, which contain fixed, inactivated tachyzoites.
6. Incubate the slides in a chamber, in the dark, at 37°C for 30 min (*see* **Note 12**).
7. Wash the slides using phosphate-buffered saline at pH 9.0 submerging the slides in a Coplin jar with a magnetic stirrer for 10 min to remove unbound antibodies (*see* **Note 13**).
8. Repeat Step 7.
9. Prepare a specific anti-mouse IgG solution by diluting it 1:40 in PBS (pH 7.2).
10. Incubate the slides again in a chamber, in the dark, at 37°C for 30 min (*see* **Note 12**).
11. Repeat the washing process as in Step 7. This time, ensure the Coplin jar is covered to protect the fluorochrome-conjugated secondary antibody from light.
12. Perform a final wash with distilled water for 10 min under the same conditions (*see* **Note 13**).
13. Apply drops of mounting fluid to the slides and place 24 × 50 × 50 mm coverslips over the mounting fluid.
14. Examine the slides under a fluorescence microscope using a 40× objective and ultraviolet light. Detect green fluorescence as an indication of positive results.
15. Set the positive cut-off point for the assay at a serum dilution of 1:10.

Positive serological results indicate that the mouse has been exposed to the parasite and has developed an immune response. Brains of seropositive mice must be analyzed by qPCR to confirm viability.

3.3.2 DNA Extraction from Mice Brain and Identification of *T. gondii*

To determine the viability of *T. gondii* in the sample, qPCR of the brain of seropositive mice is performed. For this purpose, 15 mg of brain tissue obtained in Step 8 (Sect. 3.3), should be used for DNA extraction and purification using commercial kits, following the manufacturer's protocol. The extracted DNA can be preserved by freezing until qPCR is performed.

To carry out the qPCR of brain tissue, follow Steps 1–6 as described in Sect. 3.2.

Based on the protocols described and the specific conditions of our laboratory, the threshold cycle (Ct) for sample positivity is set at

38 for protocol A and at 32 for protocol B. However, it is crucial to establish the specific limit of detection (LOD) for each laboratory during the optimization of the technique.

Positive qPCR results would confirm the viability of the parasite previously detected in the meat or meat products samples.

4 Notes

1. The sampling must be homogeneous and representative, as this step is crucial for the detection of *T. gondii* in meat and meat products.
   The amount of sample to be analyzed can be smaller or larger, but the rest of the reagents must be adjusted proportionally. Keep in mind that reducing the sample size will affect the sensitivity of parasite detection.
2. Reusable materials, such as flasks, Erlenmeyer flasks, mesh sieve, etc., used during this protocol do not need to be sterile but must be cleaned with soap and hot water (>60°C).
3. Check that the pepsin solution pH is between 1.0 and 1.2 before use and confirm that it remains below 3 after adding it to the homogenized sample to optimize enzymatic activity and ensure proper digestion. The pH can be adjusted using an HCl solution.
4. This reagent is used to neutralize pepsin and alkalize the solution. Ensure that the pH reaches approximately 6.5. This step prevents the potential destruction of bradyzoites by the acid and preserves the parasite's infectivity until it is inoculated into experimental mice.
5. To carry out the mouse bioassay, the time between digestion and inoculation into the mice must not exceed 3 h to ensure that the parasite's viability is maintained.
6. qPCR with dye-based probes, such as SYBR Green, is more sensitive than TaqMan probes. However, it requires expertise to verify specificity through the dissociation curve. This is particularly important in this case, as the target is a repetitive region that can produce signals at multiple temperatures.
7. It is recommended to perform qPCR analysis in triplicate for each sample.
8. The melting temperature (Tm) of the amplified target product is approximately 81°C. At this temperature, the double-stranded DNA denatures into single strands, confirming the specificity of the amplification. Deviations from this temperature may indicate non-specific amplification or

primer-dimer formation. However, due to the presence of contiguous repeat sequences, variations in the dissociation curve signal may occasionally occur.

9. To create a calibration curve between Ct values and DNA copies, start by preparing serial dilutions of DNA extracted from in vitro cultured *T. gondii* tachyzoites. Serial dilutions should range from 10^2 to 10^6 copies of DNA per 0.01 mL. Perform qPCR on each dilution to establish the relationship between the Ct values and the known DNA copy numbers. For each dilution, run qPCR in triplicate and calculate the average Ct values. Then, perform a linear regression analysis by plotting the Ct values (y-axis) against the log of the initial DNA copy number for each dilution (x-axis). This analysis provides the calibration curve, which can be used to quantify DNA in samples.
10. Gently pull back the plunger to check for blood or resistance, which may indicate that the needle is in a blood vessel or an organ. If blood is observed, remove the needle immediately and reposition it in a new site; if no blood is drawn, proceed with the injection.
11. The submandibular vein is located near the lower jawbone, just below the cheek. The area is easily accessible and does not require shaving.
12. To prevent severe dehydration of the slide, add water to the chamber to increase humidity during incubation.
13. Carefully use a Pasteur pipette to wash the wells to remove unbound antibodies. This can help reduce background noise in microscopic analysis.

Acknowledgments

The analytical methodologies described have been developed in our laboratory within the framework of the research projects INNPACTO IPT-2012-0189-060000 (co-financed by the Spanish Ministry of Economy and Competitiveness and the European Regional Development Fund) and RTA2014-00024-C04-02 (co-financed by the Spanish National Institute for Agricultural and Food Research and Technology, the Spanish Ministry of Economy and Competitiveness, and the European Regional Development Fund). The authors also wish to thank the Government of Aragón for its support (grant Grupo AESA).

References

1. Saadatnia G, Golkar M (2012) A review on human toxoplasmosis. Scand J Infect Dis 44:805–814
2. Ybanez RHD, Ybanez AP, Nishikawa Y (2020) Review on the current trends of toxoplasmosis serodiagnosis in humans. Front Cell Infect Microbiol 10:204
3. Vismarra A, Kramer L, Genchi M (2022) Toxoplasmosis. In: Rezaei N (ed) Encyclopedia of infection and immunity, vol. 2. Elsevier, pp 724–740
4. EFSA and ECDC (European Food Safety Authority and European Centre for Disease Prevention and Control) (2021) The European union one health 2019 zoonoses report. EFSA J 19:6406
5. Almeria S, Dubey JP (2021) Foodborne transmission of *Toxoplasma gondii* infection in the last decade. An overview. Res Vet Sci 135:371–385
6. Condoleo R, Rinaldi L, Sette S, Mezher Z (2018) Risk assessment of human toxoplasmosis associated with the consumption of pork meat in Italy. Risk Anal 38:1202–1222
7. Guo M, Dubey JP, Hill DE, Buchanan RL, Gamble HR, Jones JL, Pradhan AK (2015) Prevalence and risk factors for *Toxoplasma gondii* infection in meat animals and meat products destined for human consumption. J Food Prot 78:457–476
8. Juránková J, Basso W, Neumayerová H, Baláž V, Jánová E, Sidler X, Deplazes P, Koudela B (2014) Brain is the predilection site of *Toxoplasma gondii* in experimentally inoculated pigs as revealed by magnetic capture and real-time PCR. Food Microbiol 38:167–170
9. Opsteegh M, Prickaerts S, Frankena K, Evers EG (2011) A quantitative microbial risk assessment for meatborne *Toxoplasma gondii* infection in The Netherlands. Int J Food Microbiol 150:103–114
10. EFSA (European Food Safety Authority) (2011) Scientific report on technical specifications on harmonised epidemiological indicators for public health hazards to be covered by meat inspection of swine. EFSA J 9(10):2371
11. De Barros RAM, Torrecilhas AC, Marciano MAM, Mazuz ML, Pereira-Chioccola VL, Fux B (2022) Toxoplasmosis in human and animals around the world. Diagnosis and perspectives in the one health approach. Acta Trop 231:106432
12. Dubey JP, Hill DE, Fournet V, Hawkins-Cooper D, Cerqueira-Cézar CK, Murata FHA, Verma SK, Kwok OCH, Rani S, Fredericks J, Adams B, Jones JL, Wiegand RE, Ying Y, Guo M, Su C, Pradhan AK (2020) Low prevalence of viable *Toxoplasma gondii* in fresh, unfrozen American pasture-raised pork and lamb from retail meat stores in the United States. Food Control 109:106961
13. EFSA (European Food Safety Authority) (2018) EFSA BIOHAZ Panel (EFSA Panel on Biological Hazards), Koutsoumanis K, Allende A, Alvarez-Ordoñez A, Bolton D, Bover-Cid S, Chemaly M, Davies R, De Cesare A, Herman L, Hilbert F, Lindqvist R, Nauta M, Peixe L, Ru G, Simmons M, Skandamis P, Suffredini E, Caccio S, Chalmers R, Deplazes P, Devleesschauwer B, Innes E, Romig T, van der Giessen J, Hempen M, Van der Stede Y, Robertson L (2018) Scientific opinion on the public health risks associated with food-borne parasites. EFSA J 16:e05495
14. Chalmers RM, Robertson LJ, Dorny P, Jordan S, Kärssin A, Katzer F, La Carbona S, Lalle M, Lassen B, Mladineo I, Rozycki M, Bilska-Zajac E, Schares G, Mayer-Scholl A, Trevisan C, Tysnes K, Vasilev S, Klotz C (2020) Parasite detection in food: current status and future needs for validation. Trends Food Sci Technol 99:337–350
15. Marín-García PJ, Planas N, Llobat L (2022) *Toxoplasma gondii* in foods: prevalence, control, and safety. Foods 11:2542
16. Dubey JP (1998) Refinement of pepsin digestion method for isolation of *Toxoplasma gondii* from infected tissues. Vet Parasitol 74:75–77
17. Bayarri S, Gracia MJ, Lázaro R, Pérez-Arquillué C, Barberán M, Herrera A (2010) Determination of the viability of *Toxoplasma gondii* in cured ham using bioassay: influence of technological processing and food safety implications. J Food Prot 73:2239–2243
18. Jones JL, Dubey JP (2012) Foodborne toxoplasmosis. Clin Infect Dis 55:845–851
19. Gracia MJ, Lázaro R, Pérez-Arquillué C, Pagán R, Ramos S, Garcia JL, Bayarri S (2020) High-pressure processing (HPP) of raw and dry-cured ham from experimentally infected pigs as a potential tool for the risk control of *Toxoplasma gondii*. Innov Food Sci Emerg Technol 61:102315
20. Gracia MJ, Nieto P, Lázaro R, de Blas I, Bayarri S (2024) *Toxoplasma gondii* in Spanish commercial dry-cured meat products. Food Control 155:110079
21. Opsteegh M, Schares G, Blaga R, van der Giessen J. on behalf of the consortium (2016) Experimental studies of of *Toxoplasma gondii* in the main livestock species (GP/EFSA/BIOHAZ/2013/01) final report. EFSA Supporting Publication 13:EN-995, 161 pp

Check for updates

Chapter 9

Screening Methods for the Selection of Resistant Bacterial Variants Based on Growth Fitness in the Presence of Antimicrobial Agents

Jorge Andaluz-Arbe, Alberto Fau, Diego García-Gonzalo, Rafael Pagán, Daniel Berdejo and Diego Gómez-Lozano

Abstract

At present, foodborne diseases continue to have a high impact on Public Health, not only due to the pathogenicity of their agents, but also because of their ability to become resistant to antimicrobials, such as antibiotics, biocides, essential oils, and food preservatives. It has been shown, especially with antibiotics, that prolonged exposure to them can lead to the generation of resistant bacteria. At research level, adaptive laboratory evolution (ALE) assays and genetic editing favor the isolation of resistant bacterial populations to the antimicrobials used as stressors. Moreover, prolonged exposure can occur in natural conditions and sampling techniques in the food industry can allow the isolation, if any, of RVs. The next step is to evaluate the resistance of these evolved bacterial populations against the antimicrobial to proceed to the isolation of resistant variants (RVs) for further study or characterization. However, as it is not possible to assess the resistance of all the different strains, some methodologies should be sought as screening methods. Therefore, this chapter aims to describe a screening methodology for the isolation and selection of RVs from a heterogeneous bacterial population, both in liquid and solid media, based on growth fitness assessment in the presence of antimicrobial. The first method, known as turbidimetry, consists of randomly isolating colonies from a population and selecting RVs by comparing their growth fitness in the presence of the antimicrobial on liquid medium. Meanwhile, the second method, known as "ScanLag", compares the growth fitness in solid medium with the antimicrobial added.

Keywords Antimicrobial resistance, Bacteria, Resistant variants (RVs) selection, Growth fitness, Turbidimetry, ScanLag

1 Introduction

Nowadays, foodborne pathogens are considered a global health threat. Among foodborne pathogens, *Campylobacter* spp., *Salmonella* spp., shiga toxin-producing *Escherichia coli* (STEC), *Yersinia enterocolitica*, and *Listeria monocytogenes* are the most involved microorganisms in foodborne infections [1]. Despite the European Union's efforts to improve Food Safety, foodborne diseases and outbreaks

Marciane Magnani and Geany Targino de Souza Pedrosa (eds.), *Detection and Quantification of Bacteria, Yeast, Viruses, and Protozoan in Foods and Freshwater*, Methods and Protocols in Food Science, https://doi.org/10.1007/978-1-0716-5064-6_9,

continue to have a high impact on Public Health [2, 3]. Moreover, this problem has been aggravated by the emergence and spread of antimicrobial-resistant (AMR) bacteria along the food chain with a daunting prospect for the next years [4].

In recent years, much has been discussed about the emergence of AMR bacteria. In fact, in 2016, the United Nations General Assembly determined that the leading cause of rising AMR was due to the inappropriate use of antimicrobials in multiple sectors, including the primary sector in animal production [5]. These bacteria are present in livestock animals and, therefore, can be disseminated along the food chain. In line with the above, various authors [6–8] have demonstrated the ability of foodborne pathogens to develop resistance mechanisms to stressful environmental and processing conditions they must challenge through the food chain.

It is of interest to know the bacterial resistant variants (RVs) reservoirs. To this end, surface and food sampling are conducted in the food processing environment. However, heterogeneous populations are obtained and the evaluation of the resistance of all of them is a laborious and time-consuming process. For this reason, screening methods are needed.

On the other hand, it is also possible at the research level to determine the ability of bacteria to develop antimicrobial resistance. In this context, there are several techniques for the in vitro generation of AMR bacteria, being adaptive laboratory evolution (ALE) assays and genetic editing among the most employed. Conducting ALE assays by forcing a selective pressure on a bacterial population, enables the isolation of RVs [9]. Among the most commonly used protocols, ALE assays in which subinhibitory doses of the antimicrobial are applied during bacterial growth can be accomplished both at constant [10] and increasing doses [11], as well as with the application of lethal treatments followed by a recovery phase of the surviving bacteria [10]. As well as with the obtention of RVs by sampling, ALE assays result in heterogeneous populations where not all of them are RVs and whose selection might be complicated. For this reason, a screening method would simplify this selection task.

The implementation of screening techniques in the selection of RVs would make the RV isolation processes possible to streamline using certain parameters. One of these parameters is the growth fitness, which measures the ability of a microorganism to adapt its metabolism in a given environment [12]. Those microorganisms whose growth fitness is more adapted to a medium in the presence of an antimicrobial are more prone to have developed resistance and adaptation to it. The methods used for measuring this parameter are based on the determination of the bacterial adaptation to stress by measuring a change in growth dynamics [13, 14] in the presence of the antimicrobial agent. This change is measured by three parameters: the exponential growth rate, the lag time or lag phase and the maximum value reached (by colony count, optical density, measurement of colony area in pixels, etc.).

There are several methodologies to study the growth fitness of a microorganism, being the turbidimetry method, performed in a liquid medium (bacterial broth), one of the most employed. Moreover, there is an alternative and less used method known as "ScanLag," which is conducted in a solid medium (bacterial agar) [15].

On the one hand, the turbidimetry method is an effective technique, which consists of growing microorganisms under certain conditions in the presence of the antimicrobial, and measuring the optical density (usually between 590 and 610 nm) of the culture broth by a microplate reader [16]. It is one of the most employed methods because of its reliability and operational ease, especially if an automated microplate reader is owned.

On the other hand, the ScanLag method has been developed and validated by the authors Levin-Reisman et al. [15]. It enables lag times and growth time distributions to be assessed based on the detection of colonies on nutrient agar plates containing antimicrobial [15, 17]. Together with the method, they also designed a software to automatically analyze the images and extract quantitative parameters [18, 19], such as the time of appearance and the growth time of each colony and, afterwards, isolate the RVs directly from the same plates. The interpretation of this method follows a bimodal colony distribution where the growth curve of each colony indicates its metabolism. In this sense, colonies with higher growth in the presence of antimicrobial might be candidates for being more resistant.

Both are two methodologies to study the growth fitness of a microorganism in the presence of antimicrobial agents and act as screening methods for RVs isolation (see Fig. 1). Subsequently, a resistance assessment or genetic studies must be accomplished.

The screening by ScanLag may be more efficient and less laborious than by turbidimetry since it starts directly from the

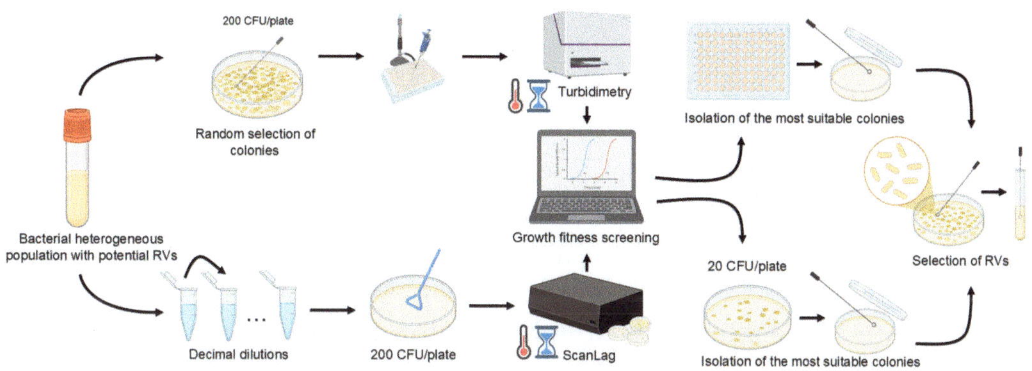

Fig. 1 Steps to be followed for the selection of RVs by evaluating their growth fitness in the presence of antimicrobials

heterogeneous population. However, ScanLag has the difficulty that, depending on the chemical properties of the antimicrobial, this might not be homogeneously distributed together with the agar.

This chapter describes a screening methodology for the isolation and selection of RVs from a heterogeneous bacterial population, both in liquid and solid media, based on growth fitness assessment in the presence of antimicrobial.

To meet the objective, two complementary methodologies are going to be described. The first method (turbidimetry) is based on the growth fitness assessment in liquid media in the presence of antimicrobial by measuring the optical density and whose comparison allows the isolation of RVs. Meanwhile, the second method, known as "ScanLag," is based on the monitoring of colony growth on nutrient agar plates with antimicrobial. Subsequently, for both methodologies, depending on the parameters obtained, the most suitable colonies will be selected to evaluate their resistance and for further phenotypic and genotypic research.

2 Material

2.1 Isolation of RVs by Turbidimetry (Growth Curves)

1. Bacterial population to study.
2. Sterile culturing tools: micropipettes, multichannel micropipette (10–100 μL), pipette tips, plastic 1.5-mL tubes, petri dishes (90 mm), 96-well microplate, inoculation loops and L-shaped spreaders.
3. Growth media: broth and agar culture media (e.g., tryptic soya agar (4%) and broth (3%) supplemented with 0.6% yeast extract).
4. Antimicrobial agents (antibiotics, biocides, food preservatives, essential oils, etc.).
5. Glass test tubes and caps.
6. 60-mL narrow-neck glass flasks.
7. Narrow neck screw caps.
8. Vortex mixer.
9. Incubator chamber with orbital shaker.
10. Microplate sealing films.
11. Sterile needles.
12. 595-nm wavelength microplate reader (advisable if it contains temperature incubation and shaking).
13. Statistical analysis and data visualization software.

2.2 Isolation of RVs by ScanLag Method

1. Bacterial population to study.
2. Sterile culturing tools: micropipettes, pipette tips, plastic 1.5-mL tubes, petri dishes (90 mm), inoculation loops and L-shaped spreaders.
3. Growth media: broth and agar culture media (e.g., tryptic soya agar (4%) and broth (3%) supplemented with 0.6% yeast extract).
4. Antimicrobial agents (antibiotics, biocides, food preservatives, essential oils, etc.).
5. Phosphate-buffered saline (PBS).
6. Glass test tubes and caps.
7. 60-mL narrow-neck glass flasks.
8. Narrow neck screw caps.
9. Vortex mixer.
10. Incubator chamber with orbital shaker.
11. Flatbed scanner with optical resolution (4800 dpi), hardware resolution (4800 × 9600 dpi) and color bit depth (48 dpi). It should be noted that the scanner must operate at the assay temperature and humidity levels.
12. Pieces of sterile black felt cloth to cover Petri dishes.
13. White plate holders designed to fit the scanner, each featuring six holes to accommodate Petri dishes.
14. Petri dishes with antimicrobial added to the agar.
15. "Scanning Manager" software application (available on the following link: https://github.com/IritLevin/ScanningManager/releases).
16. ScanLag - TimeLapse analysis software application (for MATLAB) (available on the following link: https://github.com/oferfrid/NQBMatlab/tree/V16).

3 Methods

Before describing both methodologies, it should be mentioned that the starting point is heterogeneous bacterial populations where RVs might have emerged. These bacterial populations could have been obtained through ALE assays, genetic engineering, or sampling in food industries, among other ways.

3.1 Isolation of RVs by Turbidimetry (Growth Curves)

The following Steps 1–9 are explained to obtain the initial working bacterial culture for the turbidimetry method (see Fig. 2):

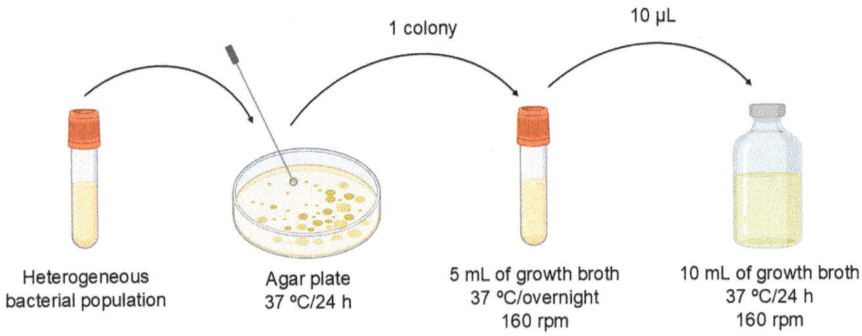

Fig. 2 Protocol for the preparation of the initial working bacterial culture for the turbidimetry method

1. From the heterogeneous bacterial population to study, inoculate and streak over the agar plate's surface with an inoculation loop.
2. Incubate agar plates for 24 h at 37°C (*see* **Note 1**).
3. Randomly select colonies and inoculate every single colony in 5 mL of growth broth in a test tube (*see* **Notes 2 and 3**).
4. Incubate overnight at 37°C in an orbital shaker at 160 rpm until a stationary phase culture is obtained.
5. Inoculate 10 μL of the overnight culture in 10 mL of growth broth in a flask.
6. Incubate at 37°C for 24 h and 160 rpm until a stationary phase culture is obtained (*see* **Note 4**).
   In order to conduct the turbidimetry assay, the following steps should be taken (continue from Step 6) (see Fig. 3):
7. Prepare the corresponding stock solutions of antimicrobial (antibiotics, biocides, food preservatives, essential oils, etc.), according to the reagent preparation (*see* **Note 5**) for latter addition to the growth medium.
8. Prepare 5-mL sterile test tubes with broth and add the corresponding volume of stock solution to obtain a concentration of antimicrobial below subinhibitory concentrations for the assay (*see* **Notes 6 and 7**). It is necessary to prepare a positive (test tube with microorganism but without antimicrobial) and a negative control (test tube with antimicrobial but without microorganism) for the assay.
9. Pour the culture broth with the added antimicrobial in a sterile Petri dish and then use a multichannel micropipette to distribute 180 μL into each microplate well. Ensure the tube is properly homogenized before pouring the content into the Petri dish (*see* **Note 8**).
10. To prepare positive controls, repeat Step 12 without adding the antimicrobial.

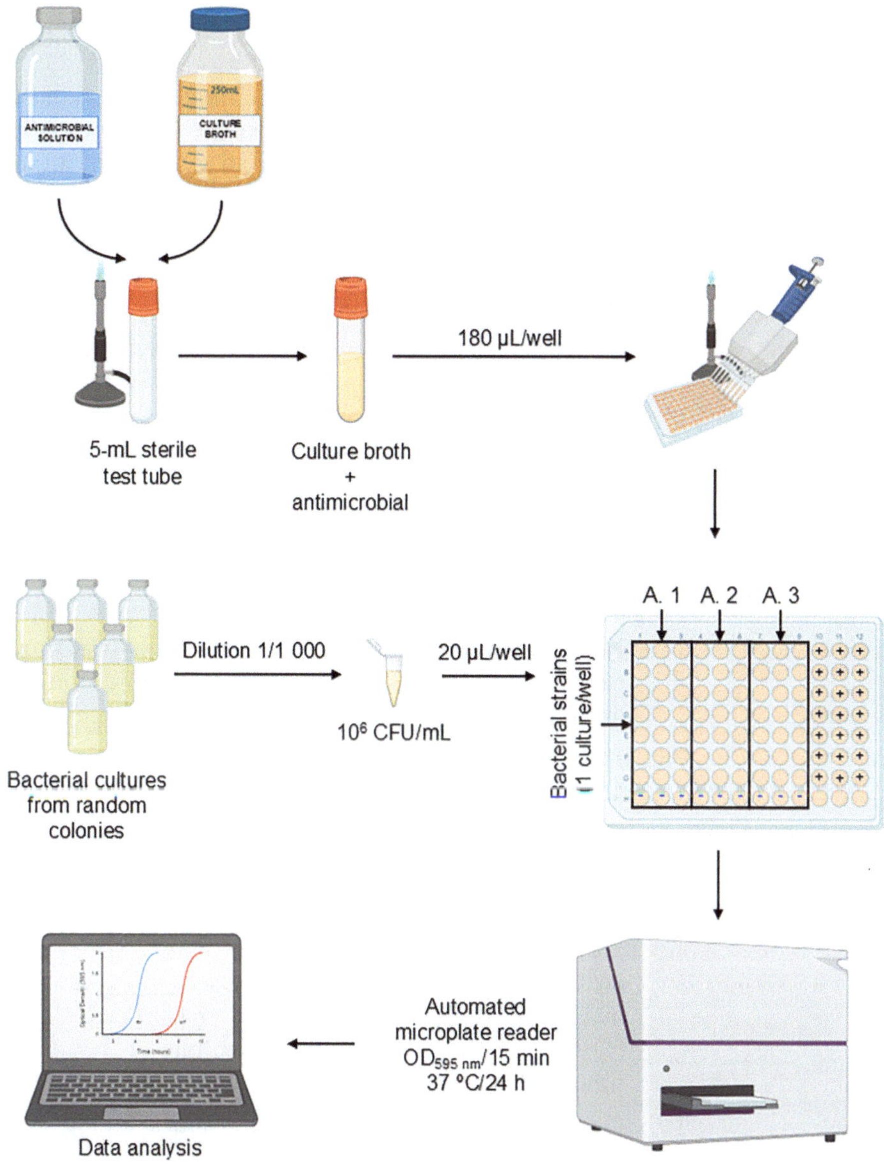

Fig. 3 Diagram of the evaluation of growth fitness by turbidimetry

11. Take the culture from Step 6 and conduct the appropriate dilutions in growth broth employing plastic 1.5-mL tubes to obtain an initial concentration of 10^5 CFU/mL in the microplate well (*see* **Notes 9 and 10**).
12. Homogenize the plastic 1.5-mL tube with the microorganism dilution and add 20 µL to each well. Homogenize up and down the total volume of the well with the aid of the micropipette (approximately 5 pipetting strokes). Skip this step when preparing negative controls.

13. Several replicates of the assay can be performed, at least in triplicate.
14. Seal the plate with a microplate sealing film to avoid contamination from the microplate reader and perform a hole in each well with a sterile needle to avoid developing an anaerobic environment in the wells.
15. Introduce the microplate inside the automated microplate reader and program the following conditions (*see* **Notes 11 and 12**):
    a. Optical density measure: 595-nm wavelength (OD_{595}).
    b. Growth temperature: 37°C.
    c. Reading time per cycle: 15 min.
    d. Shaking time for each reading: 3 s.
    e. Incubation period: 24 h (approximately 96 cycles).
16. Once the assay is accomplished, stop the incubation period and export all the data to a spreadsheet.
17. Before analyzing the data, subtract the OD_{595} at time 0 value of every measure to remove the background noise from the growth broth and the antimicrobial.
18. Display graphically the bacterial growth curves based on OD_{595} and model them using the following modified Gompertz equation [20].

$$y = A \exp\left\{ -\exp\left[\left(\mu_m \frac{e}{A} \right) (\lambda - t) + 1 \right] \right\}$$

    where y: OD_{595}; t: time (h); A: maximum value reached (OD_{595}max); μ_m: maximum growth rate (1/h); λ: lag time/time to detection (h).
19. Obtain and compare A, μ_m, and λ values among the studied colonies (see Fig. 4).
20. Select the best candidate cultures and store them in the appropriate conditions (e.g., cryovials at −80°C) for further studies.

When a studied strain shows a reduced lag time and/or a higher growth rate, it means that it is more resistant to the antimicrobial. In this sense, the strains that have shown a growth fitness in the presence of the antimicrobial (reduced lag time and higher growth rate) will be selected as the most resistant variants of the population. These RVs might be employed for their subsequent phenotypic and genotypic characterization. In the case of ALE assays and genetic editing, these values could be compared with the parental strain to

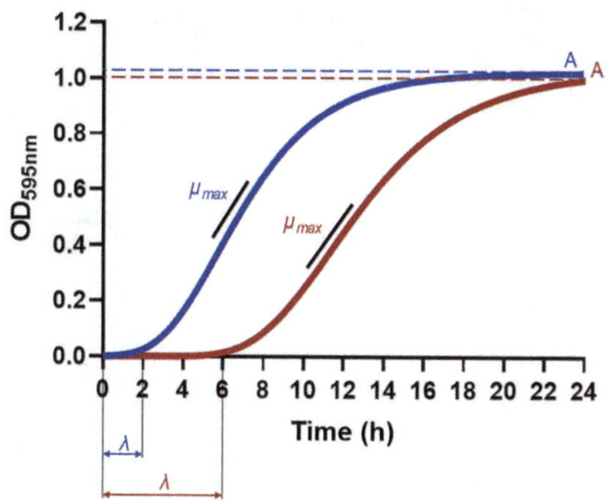

Fig. 4 Comparison example between two bacterial growth curves where the Gompertz modified values are shown: *A* (maximum value reached, OD_{595max}), μ_{max} (maximum growth rate, 1/h) and λ (lag time/time to detection, h). In the example of the graph, the studied strain, which corresponds to the blue line, should be selected as it has a lower lag time and higher growth rate in comparison to the red line

quantify their resistance increase as well as to conduct comparative genetic studies.

An important aspect that should be considered is that the assay can be performed by employing test tubes instead of a microplate (*see* **Note 13**).

3.2 Isolation of RVs by ScanLag

In order to conduct the ScanLag assay, the following steps should be taken:

1. Introduce the flatbed scanner inside the incubator at 37°C (*see* **Note 14**).
2. Place the 6-hole plate holder on the surface of the scanner (see Fig. 5).
3. Prepare the Petri dishes (*see* **Note 15**):
   a. Starting from the heterogeneous bacterial population, conduct the appropriate dilutions in PBS to obtain a bacterial concentration of 2000 CFU/mL. Afterwards, plate 0.1 mL of culture uniformly on the plate (the final bacterial concentration should be 200 CFU/plate), containing agar and antimicrobial, surface with the help of an L-shaped spreader (*see* **Note 16**).
   b. Cover the plates with a piece of sterile black felt cloth with the aim of gaining good contrast between the colonies and the scanner cover and absorb moisture. Place the lids on their corresponding Petri dishes (see Fig. 6).

Fig. 5 Scanner inside the incubator chamber with the 6-plate holders

Fig. 6 Pieces of sterile black felt cloth to cover the plates

4. Attach the plates in the holders on the scanner surface.
5. Start the "Scanning Manager":
   a. Choose the participating scanners from the list of attached scanners.
   b. Choose the number of images to be taken (repetitions) (*see* **Note 17**), the interval between subsequent images (time gap) and the time delay before starting the experiment (start after) (*see* **Note 18**).
6. Analyze the images following the instructions provided by Levin-Reisman et al. [15]:
   a. Classify the images from each scanner into separate folders.
   b. Run the following procedures employing MATLAB's command window. Find the parameters required to run these procedures defined in the software manual.
      i. To perform preprocessing, run the command "PreparePictures": In preprocessing, the images from a scanner are aligned. Each Petri dish is cropped from each sequence of images. The time at which each image was collected is retrieved from the timestamp.
      ii. To perform detection and tracking in each Petri dish separately, run the command "TLAIIPlates." This tool detects colonies and assigns them an identification number.
      iii. Perform an analysis of the extracted data: the software measures the sizes of every detected colony as a function of time. Follow the subsequent functions in MATLAB's command window with the aim of extracting the appearance distribution of the colonies:
         1. To display the analysis of a certain Petri dish alongside with the graph of the colonies size versus time, run the command "ScanLagApp":
            a. Change the current time of the plate by using the slider. The timing will also be indicated by a matching vertical line on the colonies area graph.
            b. Click on a colony to see its associated curve, and vice versa, click on a curve to find its associated colony. After identifying the colony, the phenotype can be retrieved by picking the colony from the identified Petri dish. This identification will be employed for the selection of RVs.
            c. Filter defects in the automatic analysis by choosing a colony and clicking the exclude button. When clicking exclude the number of the colony will turn yellow.

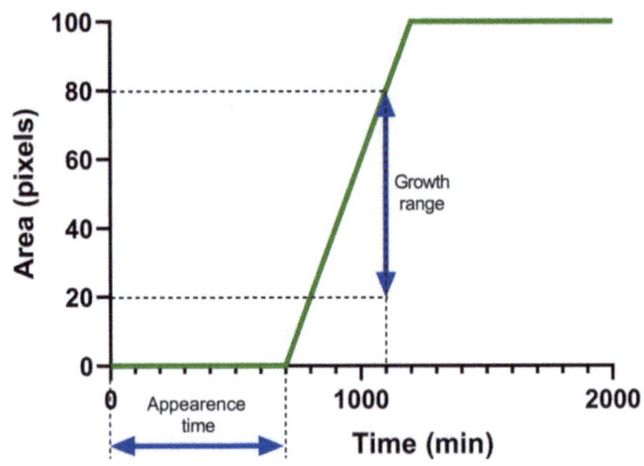

Fig. 7 Graphical representation of the area (pixels) versus time (min) of representative colonies showing the appearance time of the colonies and growth rate

2. After going through the results of the analysis of each plate, sum up the results:
   a. Create a histogram of appearance times by running "AddHistograms."
   b. Create a survival curve representation of the distribution by running "plotDeathCurve."
   c. To get the appearance time of all colonies in the experiment, run "GetAppearanceTimes."
7. To analyze the extracted data, a function was used which the colonies appearance, represented by the time the colony is detected, and the growth rate is defined as the time it takes to grow from 20 to 80 pixels (see Fig. 7). Note that the software identifies each colony with an identification number.
8. Compare the appearance time of the colonies and the growth rate values among the studied evolved populations and the parental strain (only for ALE assays and genetic editing). Since the measurements are made at least in triplicate, the mean values should be compared, and the standard deviation of the measurements should be considered.
9. As is explained in Section 3.1., select the colonies to isolate that have shown a lower appearance time and a higher growth rate as they might have a better resistance to the antimicrobial (*see* **Note 19**).
10. Store the isolates in the proper conditions for further research (e.g., cryovials at –80°C).

To sum up, both methods are good options as a screening step for the selection of RVs by understanding the growth behavior of the isolates of a heterogeneous bacterial population (*see* **Note 20**). Once the RVs have been selected, they can be phenotypically characterized (see Chapter "Phenotypic Characterization of Antimicrobial-Resistant Variants in Foodborne Pathogens"). Moreover, the genotypic characterization of these RVs will allow the study of the mechanisms responsible for their high resistance. The characterization of these RVs will provide essential information for the risk evaluation of AMR in the food chain and their impact on Public Health.

4 Notes

1. Incubation temperature and time should be modified according to the optimal growth conditions of the bacteria and the type of agar used.
2. It is recommended to select between 20 and 25 isolates per evolved population.
3. For ALE assays it is recommended to always use the original strain, to avoid the occurrence and accumulation of random mutations in the parental strain and RVs, due to their possible use in different laboratory protocols.
4. To obtain a stationary phase culture, incubation temperature and time should be modified according to the optimal growth conditions of the bacteria.
5. If the added volume of stock solution is elevated, it might dilute the growth broth. In this case, it is recommended to prepare a double-concentration broth and add the same volume as of the antimicrobial stock solution to the test tube to compensate for the dilution of the broth.
6. The concentration of antimicrobial should be adjusted to the natural antimicrobial's bacteriostatic activity.
7. It is recommended to employ half of the MIC. This recommendation is only for ALE assays, in the case of food sampling the concentration will depend on the antimicrobial and the bacterial population.
8. Organize the microplate to facilitate the interpretation of the different bacterial strains and the different antimicrobials. Take the negative and positive controls into consideration.
9. Use the formula $C_i \times V_i = C_f \times V_f$ (where C_i: initial concentration, V_i: initial volume, C_f: final concentration, V_f: final volume) to calculate the necessary dilutions. For example, in the case of *S.* Typhimurium, a 1:1000 dilution is necessary before the microorganism is added to each well. This is because the culture is diluted from an initial concentration of

10^9 CFU/mL in stationary phase to a concentration of 10^6 CFU/mL before adding to the wells.

10. This step can also be performed by employing McFarland scale.
11. The growing conditions should be programmed according to the interest growth conditions of the bacteria. The parameters "Optical density wavelength," "Reading time per cycle," "shaking time for each reading," and "incubation period" might be adjusted according to the microorganism and the cadence time interest.
12. The greater the number of measurements, the greater the accuracy of the kinetic study. However, this requires the use of more computer storage and greater difficulty in data management and analysis.
13. This approach may be necessary if an automated microplate reader with temperature control is unavailable, in which case an incubator can be used as an alternative. Furthermore, there are certain conditions which cannot be programmed in the microplate reader, such as the case of growth curves at cooling temperatures or anaerobiosis environments. In those cases, it should be noted that the reading periods might be higher due to the slower bacterial growth and the optical density must be read by a spectrophotometer.
14. It can be incubated at the desired conditions.
15. The plates must contain 20 mL of TSAYE, equivalent to 5-mm thickness.
16. For example, in the case of *S.* Typhimurium, it is necessary to conduct a 1:1,000,000 dilution and to plate 0.1 mL on the plate surface to obtain approximately 2000 CFU/mL.
17. The greater the number of measurements, the greater the accuracy of the kinetic study. However, this requires the use of more computer storage and greater difficulty in data management and analysis.
18. The usual setup is to perform 84 repetitions with an interval time of 15 min between scanning during a period of time which differs from the bacteria. For example, in the case of *E. coli*, *S.* Typhimurium and *S. aureus* it is necessary to perform the experiment for 24 h, meanwhile, in the case of *L. monocytogenes*, it is 48 h.
19. It is recommended to select around 5 isolates per evolved population.
20. Finally, it should be commented that both the turbidimetry and the ScanLag methods could be employed for the characterization of bacterial strains.

References

1. EFSA (2023) The European Union One Health 2022 zoonoses report. Sci Rep EFSA J 21:e8442
2. Pires SM, Desta BN, Mughini-Gras L, Mmbaga BT, Fayemi OE, Salvador EM, Gobena T, Majowicz SE, Hald T, Hoejskov PS, Minato Y, Devleesschauwer B (2021) Burden of foodborne diseases: think global, act local. Curr Opin Food Sci 39:152–159
3. European Food Safety Authority (2021) The European Union One Health 2019 zoonoses report. EFSA J 19(2):31–77
4. Capita R, Alonso-Calleja C (2013) Antibiotic-resistant bacteria: a challenge for the food industry. Crit Rev Food Sci Nutr 53:11–48
5. Nations U (2016) Political declaration of the high level meeting of the general assembly on antimicrobial resistance. Geneva, Switzerland
6. Berdejo D, Gayán E, Pagán E, Merino N, Campillo R, Pagán R, García-Gonzalo D (2022) Carvacrol selective pressure allows the occurrence of genetic resistant variants of *Listeria monocytogenes* EGD-e. Foods 11:3282
7. Mavri A, Smole Možina S (2013) Development of antimicrobial resistance in *Campylobacter jejuni* and *Campylobacter coli* adapted to biocides. Int J Food Microbiol 160:304–312
8. Rodríguez-Melcón C, Alonso-Calleja C, Capita R (2023) Effect of low doses of biocides on the susceptibility of *Listeria monocytogenes* and *Salmonella enterica* to various antibiotics of clinical importance. Food Control 149:109602
9. Mao EF, Lane L, Lee J, Miller JH (1997) Proliferation of mutators in A cell population. J Bacteriol 179:417–422
10. Berdejo D, Pagán E, Merino N, García-Gonzalo D, Pagán R (2021) Evolution assays for the isolation of mutant bacteria resistant to natural antimicrobials. In: Magnani M (ed) Detection and Enumeration of Bacteria, Yeast, Viruses, and Protozoan in Foods and Freshwater. Springer US, New York, NY, 65–75
11. Wong A, Rodrigue N, Kassen R (2012) Genomics of adaptation during experimental evolution of the opportunistic pathogen *Pseudomonas aeruginosa*. PLoS Genet 8:e1002928
12. Guillén S, Nadal L, Álvarez I, Mañas P, Cebrián G (2021) Impact of the resistance responses to stress conditions encountered in food and food processing environments on the virulence and growth fitness of non-typhoidal *Salmonellae*. Foods 10(3):617
13. Bochner BR (2009) Global phenotypic characterization of bacteria. FEMS Microbiol Rev 33:191–205
14. Monod J (1949) The growth of bacterial cultures. Annu Rev Microbiol 3:371–394
15. Levin-Reisman I, Fridman O, Balaban NQ (2014) ScanLag: high-throughput quantification of colony growth and lag time. J Vis Exp 89:51456
16. Levin-Reisman I, Gefen O, Fridman O, Ronin I, Shwa D, Sheftel H, Balaban NQ (2010) Automated imaging with ScanLag reveals previously undetectable bacterial growth phenotypes. Nat Methods 7:737–739
17. Balaban NQ, Merrin J, Chait R, Kowalik L, Leibler S (2004) Bacterial persistence as a phenotypic switch. Science 305:1622–1625
18. Ernebjerg M, Kishony R (2012) Distinct growth strategies of soil bacteria as revealed by large-scale colony tracking. Appl Environ Microb 78:1345–1352
19. Rotem E, Loinger A, Ronin I, Levin-Reisman I, Gabay C, Shoresh N, Biham O, Balaban NQ (2010) Regulation of phenotypic variability by a threshold-based mechanism underlies bacterial persistence. Proc Natl Acad Sci USA 107:12541–12546
20. Pagán E, López N, Sánchez A, Campillo R, Berdejo D, García-Gonzalo D, Pagán R (2024) Growth fitness, virulence, and heat tolerance of *Salmonella* Typhimurium variants resistant to food preservation methods. Int J Food Microbiol 422:110810

Chapter 10

The Use of Whole Metagenome Sequencing for Microbiome and Resistome Characterization in Food Products

Livia Balaguer, Diego García-Gonzalo, Rafael Pagán, Daniel Berdejo, Laura Espina and Natalia Merino

Abstract

Although beneficial microorganisms contribute positively to the physicochemical and sensory profile of food, the presence of spoilage or pathogenic microorganisms in food might result in undesired effects, such as organoleptic changes, food waste, economic losses, and serious health concerns. On the other hand, the extensive use of antibiotics in primary production, among other factors, has contributed to the emergence of antibiotic-resistant bacteria that might exhibit cross-resistance to other unrelated antimicrobial agents, such as biocides commonly used in the food industry. In this sense, characterizing the microbiome (collection of microorganisms) and resistome (collection of antimicrobial resistant genes) of food and food-contact surfaces is crucial for designing and implementing preventive or corrective measures to control the presence and growth of microorganisms. Culture dependent techniques have been traditionally employed to identify and quantify specific microbial groups throughout the food chain. However, their limitations and the more comprehensive insights offered by culture-independent approaches have contributed to their growing adoption in recent years. This chapter outlines the methodology for characterizing the microbiome and resistome of food and food-contact surfaces using whole metagenome sequencing (WMS).

Keywords Food microbiome, Food resistome, Antimicrobial resistance genes, Culture-independent techniques, Whole metagenome sequencing, Bioinformatics tools

1 Introduction

Microorganisms strongly determine the quality and safety of food products. Beneficial microorganisms play a crucial role in food fermentation by enhancing flavor, texture, and nutritional value, while extending shelf life through the production of antimicrobial compounds. In contrast, spoilage microorganisms are responsible for altering organoleptic characteristics of food products (off-odors, off-flavors and slime formation) [1], thus shortening their shelf-life leading to food waste and economic losses [2, 3]; while pathogenic microorganisms might be responsible for food-borne outbreaks.

Marciane Magnani and Geany Targino de Souza Pedrosa (eds.), *Detection and Quantification of Bacteria, Yeast, Viruses, and Protozoan in Foods and Freshwater*, Methods and Protocols in Food Science, https://doi.org/10.1007/978-1-0716-5064-6_10,

Additionally, the extensive use of antibiotics in primary production, formerly applied not only for treating infectious diseases, but also for prophylactic purposes and for enhancing production, has contributed to the emergence of antimicrobial-resistant (AMR) bacteria in animals. These resistant strains can be transmitted to humans through zoonotic pathways, thereby contributing to the rise of human infections difficult to treat [4]. Although there are multiple pathways of AMR dissemination, the transmission of AMR bacteria via the food chain is often underrated and warrants significant attention [5]. In this sense, identifying and understanding the dynamics of these microbial communities enables the implementation of preventative or corrective measures in food industry settings.

Culture-dependent techniques have been extensively used to determine the food-associated microbiota. These techniques rely on the isolation and subsequent identification and quantification of microorganisms [6]. While they can sometimes involve only the use of general media, as is the case for mesophile or psychrotrophes counts, they typically include selective enrichment steps (for microorganism identification), followed by selective or differential plating, and by a confirmation step through biochemical, serological, or molecular methods. Therefore, although these conventional methods are simple and inexpensive, they are time-consuming and labor-intensive, taking up to several days until results are ready [7]. Moreover, a major limitation of culturing techniques is their inability to detect fastidious microorganisms, and those in a physiologically viable but non-culturable state [6, 8], underestimating bacterial diversity as well as neglecting the interactions within the microbial communities present in the sample.

Culture-independent techniques , such as whole metagenome sequencing (WMS), are increasingly used as they offer a more accurate and greater understanding of microbial communities [9], addressing many limitations of culturing techniques. WMS involves the extraction and sequencing of the total microbial DNA content present in a sample without the need for prior amplification of specific marker genes [10]. This approach provides comprehensive information on the entire pool of genes within a sample, enabling the characterization of both its microbiome (i.e., collection of microorganisms) and its resistome (i.e., collection of antimicrobial resistance genes (ARGs)) by using specific databases [9]. However, the application of WMS requires advanced computational power and bioinformatics expertise as it generates vast amounts of high-throughput sequencing data that must be processed and analyzed using specific bioinformatics tools and reference databases. These technical demands pose significant challenges for its widespread implementation.

In this chapter, we describe the methodology for assessing the microbiome and resistome in food and food-contact surface samples through WMS. We cover all key stages, from sample collection and

DNA extraction to sequencing and simple bioinformatics analysis, with the goal of providing researchers with the necessary knowledge and tools to effectively conduct these analyses.

2 Materials

2.1 Sample Collection

1. Gloves.
2. Surface sample collection tool: sterile pre-moistened sponges.
3. Sampled area delimiting tool (sterile metal templates).
4. Sterile plastic bags.
5. Buffered peptone water (BPW).
6. Container with ice blocks for transporting samples from the food industry to the laboratory.

2.2 Sample Processing

1. Clean environment.
2. Sterile utensils for sample processing (knives and tongs).
3. Analytical balance.
4. BPW.
5. Sterile filter bags.
6. Peristaltic homogenizer (stomacher).
7. Sterile bottles.

2.3 DNA Extraction and Quantification

1. Milli-Q sterile water.
2. 15-mL sterile falcon tubes.
3. Refrigerated centrifuge capable of reaching 10,000 RCF.
4. Sterile swabs.
5. BPW.
6. Laboratory tools: Micropipettes, sterile tips, and plastic 1.5-mL tubes.
7. Microcentrifuge capable of reaching 13,000 RCF.
8. Vortex mixer.
9. DNA extraction kit: DNeasy PowerFood™ Microbial Kit or alternative kit.
10. Bead beater.
11. DNA quantification: Qubit™ 4 fluorometer, Qubit™ assay tubes, and Qubit 1× dsDNA HS Assay™, or equivalent.

2.4 Whole Metagenome Sequencing

1. Library construction kit: Nextera XT Kit or alternative kit.
2. Sequencing service provider.

2.5 Bioinformatics Analysis

1. Workstation with access to the internet and software to analyze and view sequence data.

3 Methods

3.1 Experimental Design

This study is designed to assess the microbiome and resistome of food products and food-contact surfaces using WMS. Sampling should include a representative selection of food products (e.g., raw materials, intermediate products, final products, and expired products) and food-contact surfaces (e.g., utensils, equipment, and carts). A minimum of three replicates per sample type should be collected to ensure reproducibility and to allow for statistically significant comparison between groups. Ideally, sampling should be conducted on different production days to account for potential day-to-day variability.

Moreover, the sampling strategy should be tailored to the specific objectives of the study. For example, to assess seasonal variations, samples should be collected at different times throughout the year; to evaluate changes during the product's shelf life, sampling should be performed at multiple time points during storage; or, to assess the effectiveness of cleaning and disinfection protocols, samples should be collected both before and after cleaning procedures on food-contact surfaces.

The experimental workflow—including sample collection and processing, DNA extraction and quantification, whole metagenomic sequencing, and bioinformatic analysis—is illustrated in Fig. 1.

3.2 Sample Collection

Sterile plastic bags are used to collect food samples, while food-contact surface samples are collected using sterile metal templates and pre-moistened sponges with 20 mL of sterile BPW. The sampled area depends on the sampling context. During processing, sampling 300 cm^2 of food-contact surfaces is considered adequate. In

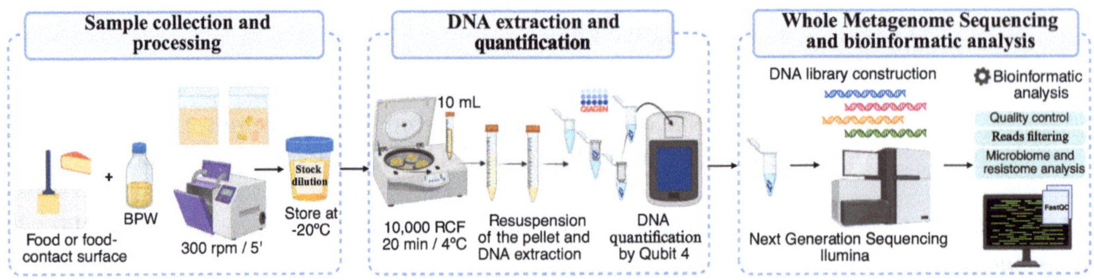

Fig. 1 Experimental workflow for the characterization of the microbiome and resistome in food products

contrast, when food-contact surfaces are sampled in a clean state—such as before the start of production, immediately after cleaning and disinfection protocols—a larger area, around 1 m^2, is generally recommended (*see* **Note 1**). Samples should be transported to the laboratory under refrigerated conditions and processed in the shortest time possible.

3.3 Sample Processing

A representative sample of food products is required for laboratory analysis, with 25 g being the standard quantity and, in many cases, the regulated amount for official testing procedures. In order to obtain the stock homogenate of each sample, 25 g of the specific food product are mixed with 225 mL of BPW in filter bags and then homogenized at 300 revolutions per minute (rpm) during 5 min using a peristaltic homogenizer. Regarding surface samples, the sponges were placed in filter bags to be homogenized. For surfaces collected during processing, 100 mL of BPW are added to filter bags. However, no BPW is added to clean surfaces to avoid DNA dilution during the homogenization step. After homogenization, the resulting dilution is transferred to a sterile bottle. This stock dilution can be used immediately or stored at −20°C for future analysis.

It is essential to ensure aseptic conditions during sample processing. Utensils used for cutting and handling food products (e.g., knives and tongs) must be thoroughly cleaned with water and soap, followed by disinfection with bleach and ethanol.

3.4 DNA Extraction and Quantification

A procedure incorporating two centrifugation steps was implemented to enhance the recovery of bacterial cells from stock homogenates, particularly when analyzing high-fat food products, as the presence of fat can hinder DNA extraction.

A 15-mL sterile Falcon is filled with 10 mL of stock homogenate, except in the case of clean food-contact surfaces, for which the entire volume of the homogenate is used for DNA extraction. This solution is centrifuged at 10,000 RCF for 20 min at 4°C (*see* **Note 2**). After centrifugation, the supernatant is discarded, the wall fat is carefully removed with a sterile swab, and the pellet is washed with 5 mL of BPW to minimize the amount of fat. Then, a second centrifugation is performed under the same conditions as described above. After discarding the supernatant and removing wall fat again, the pellet is resuspended using the first buffer (lysing reagent) of DNeasy™ PowerSoil Microbial Kit or DNeasy™ PowerFood™ Microbial Kit, and the remaining steps are carried out according to the manufacturer's instructions (*see* **Note 3**) to extract the DNA. In cases where a high concentration of host DNA, such as DNA derived from animal or plant tissues, is expected, host DNA depletion kits can be employed to reduce its abundance prior to sequencing. These kits typically function by selectively binding host DNA using functionalized magnetic particles or hybridization probes, allowing its removal through magnetic separation and thereby enriching the microbial DNA fraction.

Once extracted, the DNA concentration of each sample is measured using the Qubit™ 4 fluorometer (or any other appropriate method) (*see* **Note 4**). Until shipment to the sequencing company, DNA samples should be stored frozen to preserve their integrity. For shipping, samples are ideally transported on dry ice to maintain low temperatures and prevent DNA degradation. However, due to the inherent stability of DNA, short-term shipment at ambient temperature may also be feasible under appropriate conditions.

3.5 Whole Metagenome Sequencing

The first step involves the preparation of metagenomic libraries using library construction methods such as the Nextera XT protocol, which efficiently fragments and tags DNA to make it suitable for sequencing. Once the libraries are prepared, WMS is performed using Next Generation Sequencing platforms, particularly those from Illumina, which are widely used for metagenomics due to their high-throughput capability, accuracy, and cost-effectiveness. Illumina platforms generate short-read sequences, which enable deep sequencing coverage and comprehensive analysis of microbial communities, even in complex and diverse samples . The sequencing can be performed at an average depth of 3 Gb per sample (equivalent to 3 billion bases of DNA), providing sufficient coverage of the genetic material in the sample. For most samples, this depth enables the detection of both abundant and low-frequency genomic features within the metagenome, ensuring comprehensive analysis. The sequencing process generates paired-end reads of 151 bp each. Paired-end reads refer to sequencing both ends of a DNA fragment, producing two complementary reads. This approach improves accuracy, facilitates alignment to reference genomes, and enhances the detection of structural variations and repetitive regions.

3.6 Bioinformatic Analysis

3.6.1 Reads Pre-Processing

Raw reads often contain adapter and primer sequences, chimeric sequences, and low-quality reads. Therefore, pre-processing steps, such as trimming and quality filtering, are essential to ensure that only high-quality reads are retained for downstream analysis. The bioinformatics tool Trimmomatic v.0.33 [11] is commonly employed to remove adapter, primer and chimeric sequences and low-quality reads. Additionally, if host DNA could not be effectively reduced, raw reads may still contain host DNA sequences, which should be excluded before downstream analysis. To address this, raw reads are aligned to the host genome (*see* **Note 5**) using the bioinformatics tool bowtie2 v.2.4.5 [12], and the corresponding host sequences are then removed using Trimmomatic v.0.33 (*see* **Note 6**).

After the pre-processing steps, quality control is performed using the bioinformatics tool FastQC v.0.11.9 to ensure that the quantity and quality of the reads is sufficient for subsequent analysis. This tool provides a comprehensive overview of multiple quality metrics, including:

- General overview of reads quality, including metrics such as the number of reads, read length distribution, GC content distribution, and Phred quality scores. These metrics provide insights into the overall quality and consistency of the sequencing run, helping to identify issues like uneven read coverage or potential biases in the sequencing process.
- Base quality, including metrics such as Phred quality scores and proportion of each nucleotide at every base position. Phred scores offer an indication of the reliability of each base call (*see* **Note 7**), with high-quality sequencing data typically expected to have more than 90% of bases with a Phred score above 30. On the other hand, nucleotide composition at each base position helps identify potential issues like sequencing errors or base-calling inaccuracies, especially at the ends of reads.
- Identification of duplicate, overrepresented and k-mer sequences (short subsequences of length *k*). Duplicate sequences may indicate issues with library preparation, while overrepresented and k-mer sequences may indicate the presence of residual adapter sequences, primer dimers, or other contaminating DNA.

It is worth noting that FastQC provides both graphical and summary outputs for these parameters, offering an intuitive way to assess data quality.

3.6.2 Taxonomic Analysis

Once the trimmed and filtered metagenomic reads are selected, taxonomic assignment can be performed. This step involves classifying the reads into specific taxonomic categories (domain, kingdom, phylum, class, order, family, genus, species (*see* **Note 8**)) by comparing them to reference databases containing known genomes based on sequence similarity. Several bioinformatics tools (*see* **Note 9**) and reference databases are available for the taxonomic assignment of metagenomic data. Kraken2 v.2.1.3 [13] is commonly employed for the taxonomic classification, taking metagenomic reads in FASTA or FASTQ format (*see* **Note 10**) and the reference database in a binary format (*see* **Note 11**) as input and producing a Tab-Separated Values (TSV) file with the taxonomic assignment of each read as output. This tool uses k-mers-based classification, which involves a pre-processing step known as k-merization, where reads are broken down into short, fixed-length substrings (k-mers), followed by a classification step in which these k-mers are compared to those in a reference database (in this case, the Kraken2-Microbial Standard Database). This approach provides fast and efficient taxonomic profiling, even for large and complex metagenomic datasets, by leveraging k-mer matching, which is significantly faster and less computationally demanding than whole-sequence alignment. Finally, Bracken v.2.9 [14] can be additionally used to improve the taxonomic abundance estimates generated by Kraken2. Although Kraken2 provides accurate taxonomic classification of metagenomic reads, it does not account for genomic similarities between closely related species abundances. Bracken addresses this limitation by reestimating these abundances using probabilistic models that adjust for overlapping genomic content among species in the reference database. This refinement ensures more accurate species- and genus-level abundance estimates, particularly in complex metagenomic datasets containing near-identical species.

3.6.3 Resistome Analysis

There are several bioinformatics tools and reference databases for the identification and quantification of resistance genes in metagenomic data (*see* **Note 12**). ShortBRED v.0.9.4 [15] is commonly employed to identify and quantify antimicrobial resistance genes from trimmed and filtered metagenomic reads using the Comprehensive Antibiotic Resistance Database (CARD) as the marker database, which is widely used due to its high-quality curation and comprehensive representation of antimicrobial resistance genes [16]. ShortBRED consists of two tools: ShortBRED-Identify and ShortBRED-Quantify. First, ShortBRED-Identify reduces antibiotic resistance-related protein families from CARD into short, highly representative peptide sequences (markers) that uniquely distinguish each protein family. It takes protein families' sequences in FASTA format as input and produces the corresponding markers in FASTA format as output. Then, ShortBRED-Quantify maps metagenomic reads exclusively to those markers for identification and quantification. It takes metagenomic reads in FASTA or FASTQ format and the markers generated by ShortBRED-Identify as inputs, and produces a quantification table in TSV format detailing the number of reads mapped to each marker, corresponding to each antibiotic resistance-related protein family. This two-step process distinguishes ShortBRED from other available tools that rely on mapping metagenomic reads to full-length nucleotide or protein sequences (*see* **Note 12**), making ShortBRED faster and more specific.

See Fig. 2 for a flowchart summarizing the key bioinformatics steps, from raw read processing to taxonomic and resistome analysis.

3.6.4 Plot Generation and Statistical Analysis

The resulting taxonomic classification in the form of a TSV file (*see* **Note 13**) should be imported to the R 4.3.2 environment for further data analysis. The relative abundance of the most abundant taxa is visualized using cumulative bar plots, with taxa representing less than 5% across all communities grouped into the "Others" category. Statistical differences in taxonomic relative abundances across different communities are assessed through the Wilcoxon test using

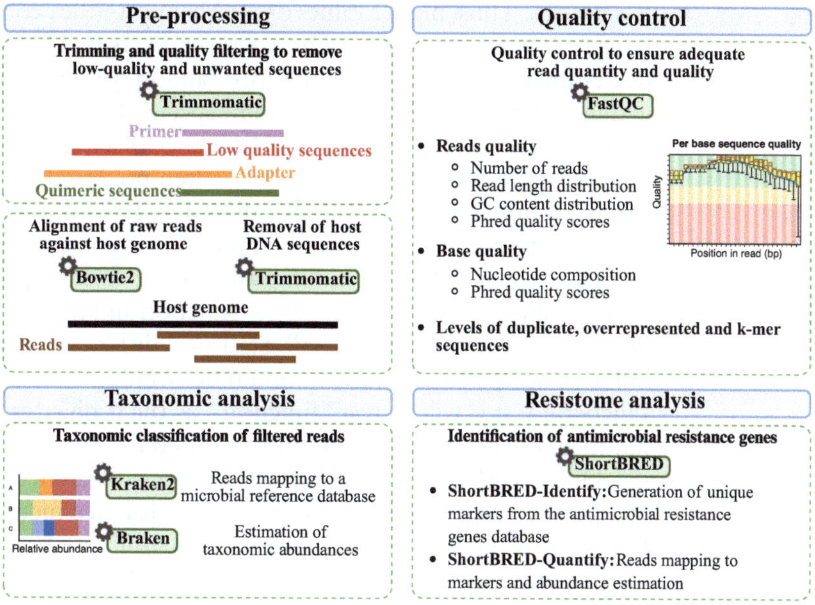

Fig. 2 Bioinformatic workflow for reads preprocessing, quality control, and subsequent taxonomic and resistome analysis

the function "wilcox.test". Similarly, the resistome composition, also provided in TSV format, is imported into the R 4.3.2 environment for further analysis (*see* **Note 13**). The abundance of the most abundant ARGs associated with resistance to specific antibiotic families is visualized using cumulative bar plots. ARGs abundance is expressed in RPKM (Reads per Kilobase of transcript per Million mapped reads), a normalization method that adjusts the number of reads mapped to an ARG for both the length of the ARG and the total number of reads in a sequencing run. This normalization technique allows for a more accurate comparison of relative abundance between ARGs, even when they differ in length or originate from samples with varying sequencing depths (*see* **Note 14**).

In addition to the representation of microbiome and resistome composition, the *vegan* v2.5.7 R package [17] was used to analyze α-diversity and β-diversity.

On the one hand, α-diversity measures diversity within a single community, encompassing both species richness (the number of species present) and evenness (the distribution of species relative abundances). Common indices include Observed Species, Shannon, Simpson, and Inverse Simpson. The Observed Species-index measures the number of species richness. Higher values correspond to a greater number of species, thus reflecting increased richness. In contrast, both the Shannon- and Simpson- indices consider not only richness but also evenness. The Shannon-index assigns higher values when there is both a high number of species and a more even distribution of their abundances, while lower values indicate a community dominated by a few species. Thus, higher values of the Shannon-index correspond to greater diversity, and lower values to lower diversity. The Simpson-index estimates the probability that two randomly selected individuals within a community belong to the same species. Higher values indicate lower diversity, as they reflect the dominance of a few species. Consequently, the Inverse Simpson-index is often preferred, as it provides a more intuitive interpretation of diversity. The number of species (Observed Species-index) was computed using the "specnumber" function, whereas Shannon, Simpson and Inverse Simpson indexes were computed using the "diversity" function and the metrics "shannon", "simpson", "invsimpson". Statistical differences in α-diversity index values across different communities were assessed through the Wilcoxon test ("wilcox.test" function), as the α-diversity values did not meet the normality assumption required for parametric tests.

On the other hand, β-diversity assesses differences in species composition between communities using distance metrics, such as Bray-Curtis distance, which measures dissimilarity based on both the presence/absence of species and their relative abundances. To visualize these differences, dimensionality reduction techniques like Principal Coordinates Analysis (PCoA) are commonly applied. In a PCoA plot, each point represents a different community, and the distance between points, determined by the distance metric, reflects the dissimilarity between those communities. Points that are close to each other indicate similar community compositions, while distant points indicate greater dissimilarity. Groupings of samples in the plot may be influenced by underlying factors, which can help identify patterns or correlations between community composition and these factors. Here, the Bray–Curtis distance was calculated using the "vegdist" function and the metric "bray" and visualized in a PCoA plot. Finally, statistical differences in β-diversity among different communities were assessed using Permutational Multivariate Analysis of Variance (PERMANOVA) with the "adonis" function. This non-parametric test, well-suited for microbiome data, evaluates differences between communities based on a distance matrix, such as Bray-Curtis distance.

3.6.5 Deposition of Sequencing Data in Public Repositories

Many scientific journals and funding agencies require data submission to promote compliance with the FAIR principles (Findable, Accessible, Interoperable, and Reusable), which aim to enhance the accessibility and long-term utility of scientific data, while also supporting broader commitments to scientific transparency and ethical research practices. In this context, raw sequencing data, typically quality-controlled raw reads in FASTQ format, should be deposited in publicly accessible repositories, such as the National Center for Biotechnology Information (NCBI) Sequence Read Archive (SRA) or European Nucleotide Archive (ENA) (*see* **Note 15**). Data can be submitted through the NCBI SRA Submission Portal, ENA Webin Portal or ENA Webin-CLI, but the command-line submission is also an option. It is important to note that both databases are synchronized, ensuring consistency and accessibility across platforms.

In addition to sequencing data, these repositories require metadata that provide a clear overview of the project. For example, the NCBI SRA Submission Portal requires the completion of four sections: Project, Sample, Library, and Files. The Project section includes detailed information about the study, such as the project title, a description of the research, funding sources and related publications. The Sample section provides essential details about each biological sample, including sample type, collection date and experimental conditions. The Library section specifies the sequencing method, including the sequencing platform used and the library preparation protocol. Finally, the Files section requires the uploading of raw sequencing data files and linking them to the corresponding metadata to ensure full contextualization.

4 Notes

1. Larger areas are sampled in clean surfaces, where the expected microbial load is low. Conversely, such extensive sampling is not necessary for used surfaces sampled during processing due to the typically higher microbial presence. Moreover, the choice of sampling tool depends on several factors, including the area size, its accessibility, and the type of surface. Swabs are generally employed for small, inaccessible areas, while gauzes or sponges are more suitable for large, accessible areas. Sampling tools can be dry, or premoistened with the diluent to improve the recovery of microorganisms.
2. To achieve optimal separation of the fat layer, centrifugation should be performed at low temperatures; therefore, using a centrifuge pre-refrigerated to 4°C is recommended.
3. To obtain an optimal final DNA concentration, the DNA is initially eluted in 25 μL of elution buffer. This same volume is then reapplied to the column for a second elution to maximize recovery.
4. The minimum DNA concentration required for metagenomic sequencing depends on the specifications of the selected sequencing company, which may vary according to the sequencing platform and library preparation protocol. It is therefore advisable to consult the sequencing company's guidelines in advance to ensure compliance with quality and quantity requirements.
5. Genomes are available for download from the NCBI (National Center for Biotechnology Information) website, which integrates various databases, including GenBank. The search tool can be used to locate specific genomes by genome identifier or taxonomic classification.

6. Depending on the sample type, host DNA may account for more than 90% of total sequences, limiting the proportion of microbial reads for analysis.
7. The Phred score indicates the probability that a given base call is incorrect. A Phred score of 10 corresponds to an error probability of 1 in 10 (10%); a score of 20 to 1 in 100 (1%); and a score of 30 to 1 in 1000 (0.1%). Therefore, higher Phred scores reflect greater base-calling accuracy and lower error probabilities.
8. It is important to follow taxonomic nomenclature conventions when reporting microbial data. Kingdom, phylum, class, and order begin with a capital letter but are not italicized (e.g., order Enterobacterales), whereas family, genus, and species begin with a capital letter and are italicized (e.g., family *Enterobacteriaceae*).
9. The most used bioinformatics tools for taxonomic classification of metagenomic data include Kraken, MetaPhlAn, Centrifuge, DIAMOND, and MEGABLAST.
10. FASTA is a plain-text format that contains a sequence identifies followed by the sequence itself, whereas FASTQ is an extension of FASTA that incorporates additional quality score information for each base in the sequence.
11. The binary format is optimized for efficient storage and fast access. Unlike plain-text formats, it reduces computational burden, enhances speed, and efficiently manages large-scale datasets, minimizing memory usage and improving performance. This format allows Kraken2 to perform taxonomic classification swiftly.
12. The most widely used bioinformatics tools for resistome analysis include ShortBRED, DeepARG, ResFinder and AMR++. The most commonly used reference databases are CARD, ResFinder, ARG-ANNOT and MEGARes.
13. Ensure that the TSV files generated by the taxonomic and resistome classification pipelines are formatted correctly (e.g., tab-delimited, with appropriate headers) before importing into R, to avoid parsing errors.
14. Alternative normalization methods, such as TPM (Transcripts Per Million), CPM (Counts Per Million) or normalization relative to 16S rRNA gene abundance, may also be considered depending on the study objectives or disciplinary conventions.
15. It is possible to request an embargo period when submitting data to public repositories, delaying public access to the data until publication or a specified release date.

References

1. Gram L, Ravn L, Rasch M, Bruhn JB, Christensen AB, Givskov M (2002) Food spoilage—interactions between food spoilage bacteria. Int J Food Microbiol 78:79–97
2. Ishangulyyev R, Kim S, Lee SH (2019) Understanding food loss and waste-why are we losing and wasting food?. Foods 8:297
3. Nychas G-JE, Panagou EZ, Mohareb F (2016) Novel approaches for food safety management and communication. Curr Opin Food Sci 12:13–20
4. WHO (2012) The evolving threat of antimicrobial resistance: options for action. World Health Organization, Geneva
5. EFSA (2008) Scientific Opinion of the Panel on Biological Hazards on a request from the European Food Safety Authority on foodborne antimicrobial resistance as a biological hazard. EFSA J 765:1–87
6. Mayo B, Rachid CT, Alegría A, Leite AM, Peixoto RS, Delgado S (2014) Impact of next generation sequencing techniques in food microbiology. Curr Genomics 15:293
7. Mandal P, Biswas A, Choi K, Pal UK (2011) Methods for rapid detection of foodborne pathogens: an overview. Am J Food Technol 6:87–102
8. Forbes JD, Knox NC, Ronholm J, Pagotto F, Reimer A (2017) Metagenomics: the next culture-independent game changer. Front Microbiol 8:1069
9. Jagadeesan B, Gerner-Smidt P, Allard MW, Leuillet S, Winkler A, Xiao Y, Chaffron S, Van Der Vossen J, Tang S, Katase M, McClure P, Kimura B, Ching Chai L, Chapman J, Grant K (2019) The use of next generation sequencing for improving food safety: Translation into practice. Food Microbiol 79:96–115
10. Cao Y, Fanning S, Proos S, Jordan K, Srikumar SA (2017) A Review on the applications of next generation sequencing technologies as applied to food-related microbiome studies. Front Microbiol 8:1829
11. Bolger AM, Lohse M, Usadel B (2014) Trimmomatic: a flexible trimmer for Illumina sequence data. Bioinformatics 30:2114–2120
12. Langmead B, Salzberg SL (2012) Fast gapped-read alignment with Bowtie 2. Nat Methods 9:357–359. https://doi.org/10.1038/nmeth.1923
13. Wood DE, Lu J, Langmead B (2019) Improved metagenomic analysis with Kraken 2. Genome Biol 20: 257
14. Lu J, Breitwieser FP, Thielen P, Salzberg SL (2017) Bracken: estimating species abundance in metagenomics data. PeerJ Comput Sci 3:104
15. Kaminski J, Gibson MK, Franzosa EA, Segata N, Dantas G, Huttenhower C (2015) High-specificity targeted functional profiling in microbial communities with ShortBRED. PLoS Comput Biol 11:1004557
16. Alcock BP, Huynh W, Chalil R, Smith KW, Raphenya AR, Wlodarski MA, Edalatmand A, Petkau A, Syed SA, Tsang KK, Baker SJC, Dave M, McCarthy MC, Mukiri KM, Nasir JA, Golbon B, Imtiaz H, Jiang X, Kaur K, Kwong M, Liang ZC, Niu KC, Shan P, Yang JYJ, Gray KL, Hoad GR, Jia B, Bhando T, Carfrae LA, Farha MA, French S, Gordzevich R, Rachwalski K, Tu MM, Bordeleau E, Dooley D, Griffiths E, Zubyk HL, Brown ED, Maguire F, Beiko RG, Hsiao WWL, Brinkman FSL, Van Domselaar G, McArthur AG (2023) Expanded curation, support for machine learning, and resistome prediction at the Comprehensive Antibiotic Resistance Database. Nucleic Acids Res 51:690–699
17. Oksanen J, Blanchet FG, Friendly M (2022) Vegan: community ecology package. R package version 2.5-7

References

Chapter 11

Molecular Detection of Hepatitis E Virus in Pork Meat

Meriane Demoliner, Juliana Schons Gularte and Fernando Rosado Spilki

Abstract

Hepatitis E virus (HEV), especially genotypes 3 and 4, is an emerging zoonotic pathogens commonly associated with foodborne transmission, particularly through the consumption of contaminated pork and pork products. This chapter presents a methodology for the molecular detection of HEV in pork meat samples, describing the preparation of raw meat samples for analysis, nucleic acid extraction using magnetic bead technology, cDNA synthesis, RT-nested PCR, and electrophoresis.

Keywords Hepatitis E, Foodborne, Pork meat, Zoonotic transmission

1 Introduction

Hepatitis E virus (HEV) is a significant foodborne pathogen, primarily linked to the consumption of pork and pork products [1, 2]. Foodborne cases are predominantly caused by HEV genotype 3, with a smaller contribution from genotype 4. While domestic pigs are the main source of HEV, game animals such as wild boar, deer, and rabbits also act as reservoirs [3, 4]. Due to HEV's global impact and zoonotic transmission through pork, monitoring HEV in pork is critical for identifying contamination and preventing outbreaks.

For the molecular detection of HEV in pork meat, we selected the magnetic beads method for nucleic acid extraction and purification due to its high binding capacity, efficient elution in small volumes, and versatility for both low and high-throughput applications. The use of RT-nested PCR enhances sensitivity [5]. To target multiple HEV strains, we used degenerate oligonucleotides that amplify ORF1 of HEV, which is phylogenetically distinct across HEV species [6, 7].

This introduction is kept brief to avoid repetition of broader HEV background information covered in the previous chapter, focusing here specifically on methodologies for the molecular detection of HEV in pork meat.

Marciane Magnani and Geany Targino de Souza Pedrosa (eds.), *Detection and Quantification of Bacteria, Yeast, Viruses, and Protozoan in Foods and Freshwater*, Methods and Protocols in Food Science, https://doi.org/10.1007/978-1-0716-5064-6_11,

2 Materials

2.1 Preparation of Raw Meat Samples

1. Sterile scalpel blade.
2. Petri dish or another sterile surface.
3. Sterile 2 mL microtubes.
4. Sterile saline solution.
5. Vortex.
6. Centrifuge.

2.2 Nucleic Acid Extraction

1. 1.5 mL or 2.0 mL DNAse-free RNAse microtubes.
2. DNAse-free RNAse-free barrier tips: 10 μL, 20 μL, 100 μL, 200 μL, and 1 mL (in racks).
3. King Fisher Deep Well extraction plates for 24 samples.
4. King Fisher Duo 12 Tip Comb.
5. MagMAX™ CORE Nucleic Acid Purification Kit (Applied Biosystems, Ref: A32700).
6. Vortex mixer.
7. Minicentrifuge (for quick spins).
8. Laminar flow.
9. Sample storage refrigerator.
10. King Fisher Duo Prime automated extraction system (Thermo Scientific™).

2.3 cDNA Synthesis

1. GoScript™ Reverse Transcription Mix, Random Primers (Promega Corporation, USA).
2. Nuclease-free, low retention qPCR-compatible reaction tubes or plates.
3. Pipettes and sterile, aerosol-resistant tips.
4. Experimental and reference RNA samples (previously extracted RNA).
5. Thermal cycler or 25°C, 42°C, and 70°C controlled-temperature heat blocks.
6. Ice or 4°C cooling blocks.
7. Vortex mixer.
8. Minicentrifuge.
9. Laminar flow.
10. Sample storage refrigerator.

2.4 Nested PCR

1. GoTaq® Colorless Master Mix.
2. Primers (see in Table 1).

Table 1
Oligonucleotide used to perform Nested-PCR

Oligonucleotides	Sequence	Amplicon length (bp)
HEVORF1con-s1	5′-CTG GCA TYA CY CTA CTG CYA TTG AGC-3′	418
HEVORF1con-a1	5′-CCA TCR ARR CAG TAA GTG CGG TC-3′	
HEVORF1con-s2	5′-CTG CCY TKG CGA ATG CTG TGG-3′	
HEVORF1con-a2	5′-GGC AGW RTA CCA RCG CTG AAC ATC-3′	287

3. Nuclease-free water.
4. Thermal cycler (PCR machine).
5. Micropipettes.
6. Tips.
7. PCR tubes 0.2 mL.
8. PCR cabinet.
9. Vortex mixer.
10. Microcentrifuge tube spinner.

2.5 Electrophoresis

1. Agarose.
2. 1× TBE buffer.
3. Ethidium bromide.
4. DNA ladder.
5. Loading dye.
6. Electrophoresis system.
7. UV transilluminator.
8. Micropipettes.
9. Tips.
10. Flask/Beaker.

3 Methods

3.1 Preparation of Raw Meat Samples Procedure

1. Take approximately 1 g of raw meat and place it on a Petri dish.
2. Use a sterile scalpel blade to chop the meat sample until it reaches a homogeneous, pâté-like consistency.
3. Transfer the chopped meat to a sterile 2 mL microtube.

4. Add 1 mL of sterile saline solution to the microtube containing the meat sample.
5. Close the microtube and vortex the sample for approximately 30 s, or until the mixture becomes homogeneous.
6. Incubate the sample for 1 h at 4°C.
7. Centrifuge the sample at 4000–5000 × *g* for 5 min to sediment the solids.
8. Collect the supernatant, which is now ready for nucleic acid extraction.

3.2 Nucleic Acid Extraction Procedure

Viral RNA extraction is performed by a commercial kit MagMAX™ CORE Nucleic Acid Purification Kit (Applied biosystems™, Thermo Fisher Scientific, Waltham, MA, USA), using automated equipment KingFisher™ Duo Prime (Thermo Fisher Scientific™), following the manufacturer's protocol. (https://assets.thermofisher.com/TFS-Assets/LSG/manuals/MAN0015944_MagMAXCORE_NA_Kit_UG.pdf).

3.2.1 Set Up the Processing Plates

1. Plate setup: KingFisher™ Duo Prime
2. (Optional) To prevent evaporation and contamination, cover the prepared processing plates with sealing foil until they are loaded into the instrument.

Plate position	Plate type	Reagent	Volume per well
B and F	Deep well 96 plate	MagMAX™ CORE Wash Solution 1	500 μL
C and G	Deep well 96 plate	MagMAX™ CORE Wash Solution 2	500 μL
D and H	Deep well 96 plate	MagMAX™ CORE Elution Buffer	90 μL
D and H	Deep well 96 plate	Place a tip comb on the plate.	

3.2.2 Prepare Bead/PK Mix

1. We recommend that you prepare a new Bead/PK Mix for each processing run. If necessary, you can store Bead/PK Mix at 4°C for up to 1 week.
2. Vortex the MagMAX™ CORE Magnetic Beads thoroughly to ensure that the beads are fully resuspended.
3. Combine the following components for the required number of samples, plus 10% overage (recommended).

Component	Volume per sample
MagMAX™ CORE Magnetic Beads	20 μL
MagMAX™ CORE Proteinase K	10 μL
Total Bead/PK Mix	30 μL

3.2.3 Prepare Lysis/Binding Solution

1. Combine the following components for the required number of samples plus 10% overage.

Component	Volume per sample
MagMAX™ CORE Lysis Solution	350 μL
MagMAX™ CORE Binding Solution	350 μL
Total Lysis/Binding Solution	700 μL

2. Mix by inverting the tube or bottle at least 10×.
3. (Optional) Store Lysis/Binding Solution at room temperature for up to 24 h.

3.2.4 Prepare the Sample

1. Prepare samples according to the sample type described above.

3.2.5 Combine the Sample with Bead/PK Mix and Lysis/Binding Solution

1. Invert the tube of Bead/PK Mix several times to resuspend the beads, then add 30 μL of the Bead/PK Mix to the required wells in the A and E plate positions.
2. Transfer the appropriate volume of each prepared sample (200 μL) to a well with Bead/PK Mix.
3. Mix the sample with the Bead/PK Mix for 2 min at room temperature according to your mixing method.
4. Add 700 μL of Lysis/Binding Solution to each sample-containing well.

   Immediately proceed to load the samples into the instrument.

3.2.6 Process Samples on the Instrument

1. Select the appropriate script on the instrument (see "Download and install the script" on MagMAX™ CORE Nucleic Acid Purification Kit User Guide).
2. Start the run, then load the prepared plates in the appropriate positions when prompted by the instrument.
3. Store purified nucleic acid on ice for immediate use, or at −80°C for long-term storage.

3.2.7 cDNA Synthesis Procedure

After RNA extraction, reverse transcription is conducted using the GoScript™ Reverse Transcription Mix, Random Primers (Promega Corporation, USA) following the manufacturer's instructions (https://www.promega.com.br/-/media/files/resources/protocols/product-information-sheets/a/9pia280-goscript-reverse-transcription-mix_random-primers.pdf?rev=7eb3c166c37c41a38d8e915f50a4111d&sc_lang=en).

3.2.8 First-Strand cDNA Synthesis

1. Thaw the GoScript™ Reverse Transcription Mix components on ice, mix gently, and centrifuge briefly.
2. Prepare 10 μL of GoScript™ Reverse Transcription Mix for each reaction on ice, combining the components in the following order. Mix gently after each addition:

Component	Volume
Nuclease-free water	4 μL
GoScript™ Reaction Buffer, Random Primer	4 μL
GoScript™ Enzyme Mix	2 μL
Final volume	10 μL

3. Gently mix the GoScript™ Reverse Transcription Mix by pipetting or vortexing.
4. Optional: To denature RNA with high degrees of secondary structure (>60% GC content), incubate RNA samples at 70°C for 5 min, then immediately chill in ice water for at least 5 min. Centrifuge in a microcentrifuge for 5 s.
5. Combine 10 μL of GoScript™ Reverse Transcription Mix and up to 10 μL of RNA sample in each reaction tube or well of a reaction plate on ice as shown below.

Component	Volume
GoScript™ Reverse Transcription Mix	10 μL
Experimental or reference RNA	10 μL[a]
Nuclease-free water	To a final volume of 20 μL[b]
Final volume	20 μL

[a]If using a sample XE "Sample" volume of less than 10 μL, adjust the volume by adding nuclease-free water to reach a final reaction XE "Reaction" volume of 20 μL
[b]You will only use water if the sample volume is less than 10 μL

6. Mix each reverse transcription reaction by pipetting. Close or seal the reaction tubes or wells.
7. Incubate the reverse transcription reactions in controlled-temperature blocks or in a programmed thermal cycler.

Step	Temperature	Time	Number of cycles
Anneal primer[a]	25°C	5 min	1 cycle
Extension	42°C[b]	60 min	1 cycle
Inactivation	70°C[c]	15 min	1 cycle
Hold	4°C	∞	1 cycle

[a]Optional
[b]The extension temperature may be optimized between 37 and 55°C
[c]Higher temperatures up to 95°C for 5 min may be used for the inactivation step

8. Store the reaction products at 4°C in the reaction tubes or wells for immediate analysis (up to 24 h) or at −80°C for long-term storage.

3.2.9 Nested PCR Procedure

1. Prepare the first master mix by combining the GoTaq® Colorless Master Mix, 2×, primers, and nuclease-free water in a PCR tube, keeping it on ice.
2. For a final reaction volume of 50 μL, use the following proportions:

Components	Volume per reaction (μL)
GoTaq® Colorless Master Mix, 2×	25
Forward primer (20 μM stock)[a]	1
Reverse primer (20 μM stock)[b]	1
Nuclease-free water	18

[a]HEVORF1con-s1
[b]HEVORF1con-a1

3. Add 5 μL of DNA template (cDNA product) to the reaction mix.
4. Mix the reaction gently and spin briefly to bring the contents to the bottom of the tube.
5. Place the reaction tube into the thermal cycler and start the PCR program:

Step	Temperature (°C)	Time	Cycles
Denaturation	95	5 min	1
Denaturation, annealing, and extension	95	30 s	45
	59	1 min	
	72	1 min	
Final extension	72	7 min	1
Hold	4	Hold	∞

3.2.10 Second-Round (Nested) PCR

The second-round (nested) PCR was performed using the same reaction mixture composition, final volume, and thermal cycling conditions described above, except for the primer pair and the template.

The primers HEVORF1con-s2 (forward) and HEVORF1con-a2 (reverse) were used in place of HEVORF1con-s1 and HEVORF-1con-a1.For each reaction, 5 μL of the amplicon obtained from the first-round PCR was used as the template instead of cDNA.

All other reaction components, concentrations, and PCR conditions remained unchanged.

3.2.11 Electrophoresis

1. In a Flask/Beaker add 2 g of agarose powder and 100 mL of 1× TBE buffer.
2. Heat the mixture in a microwave in short bursts (~20–30 s) until the agarose is completely dissolved.
3. For each mL of TBE buffer use 0.1 μL of ethidium bromide, in this case to 100 μL used 10 μL
4. Pour the agarose solution into the gel casting tray and insert the comb to create wells. At room temperature wait until the gel solidifies.
5. Place the solidified gel into the electrophoresis tank and fill the tank with 1× TBE buffer until the gel is submerged.
6. Carefully load the DNA ladder (molecular weight marker) by mixing it with loading dye as recommended by the manufacturer.
7. Carefully load the 10 μL of the mixture of 10 μL the PCR products with 2 μL of 6× loading dye.
8. Connect the gel apparatus to the power supply and run the gel at 30 volts for 10 min, following 70 volts for 60 min.
9. After the run is complete, remove the gel and view it using a UV transilluminator to visualize the DNA bands.
10. Compare the PCR product bands with the positive control and the DNA ladder to determine the size of the amplified products

4 Notes

4.1 Preparation of Raw Meat Samples

1. Handle the supernatant carefully to avoid cross-contamination between samples.

4.2 Nucleic Acid Extraction

1. Ensure that all steps are performed with DNAse and RNAse-free consumables.
2. Check your MagMAX™ CORE kit to ensure all components (lysis buffer, wash buffers, elution buffer, magnetic beads, etc.) are available before starting the extraction process.

4.3 Nested PCR

1. Primer dimers or non-specific products may appear. Optimize primer concentrations and annealing temperature, consider using a gradient PCR to find the best annealing temperature.
2. Nested PCR is sensitive to contamination, which can lead to false positives. Always use negative control and handle samples carefully to avoid contamination.
3. It's possible that the sample contains inhibitors that affect the PCR reactions. Performing serial dilutions of the sample or cDNA may significantly reduce this inhibition and improve amplification efficiency.
4. Additional sensitivity can be achieved by employing a touch-down PCR approach, which optimizes amplification by gradually decreasing the annealing temperature during the initial cycles.

4.4 Electrophoresis

1. Ethidium bromide is a hazardous chemical; handle it with care, using appropriate PPE.

4.5 Positives Samples

Since HEV is divided into multiple genotypes, sequencing and characterization of positive samples are crucial for identifying the specific HEV strains in circulation

Acknowledgments

Our studies are funded by grants from Fapergs (n° 23/2551-0002221-4) and CNPq (n° 405786/2022-0).

References

1. World Health Organization (2023) Hepatitis E, https://www.who.int/news-room/fact-sheets/detail/hepatitis-e (accessed 30 Aug 2024)
2. Teshale E (2024) Hepatitis E. CDC Yellow Book 2024: Travel-Associated Infections & Diseases, https://wwwnc.cdc.gov/travel/yellowbook/2024/infections-diseases/hepatitis-e#:~:text=Epidemiology,Asia%2C%20and%20tropical%20East%20Asia (accessed 30 Aug 2024)
3. Ricci A, Allende A, Bolton D, Chemaly M, Davies R, Fernandez Escamez PS, Herman L, Koutsoumanis K, Lindqvist R, Nørrung B, Robertson L, Ru G, Sanaa M, Simmons M, Skandamis P, Snary E, Speybroeck N, Ter Kuile B, Threlfall J, Wahlström H, Di Bartolo I, Johne R, Pavio N, Rutjes S, van der Poel W, Vasickova P, Hempen M, Messens W, Rizzi V, Latronico F, Girones R (2017) Public health risks associated with hepatitis E virus (HEV) as a food-borne pathogen. EFSA J 15:e04886
4. Pavio N, Doceul V, Bagdassarian E, Johne R (2017) Recent knowledge on hepatitis E virus in Suidae reservoirs and transmission routes to human. Vet Res 48:1–14
5. Kittigul L, Ekchaloemkiet S, Utrarachkij F, Siripanichgon K, Sujirarat D, Pungchitton S, Boonthum A (2005) An efficient virus concentration method and RT-nested PCR for detection of rotaviruses in environmental water samples. J Virol Methods 124:117–122
6. Erker JC, Desai SM, Mushahwar IK (1999) Rapid detection of Hepatitis E virus RNA by reverse transcription-polymerase chain reaction using universal oligonucleotide primers. J Virol Methods 81:109–113
7. Purdy MA, Harrison TJ, Jameel S, Meng XJ, Okamoto H, Van der Poel WHM (2022) ICTV Report Consortium. ICTV virus taxonomy profile: Hepeviridae. J Gen Virol 98:2645–2646

Chapter 12

Molecular Detection of Hepatitis E Virus in Water

Juliana Schons Gularte, Meriane Demoliner and Fernando Rosado Spilki

Abstract

Hepatitis E virus (HEV) is a significant cause of acute hepatitis in humans and an emerging zoonotic disease. HEV genotypes 1 and 2 primarily affect humans, and HEV-3 and HEV-4 are mainly transmitted from zoonotic sources. Transmission occurs through contaminated water, undercooked meat, and, in some cases, via blood transfusions or organ transplants. Globally, HEV is responsible for approximately 20 million infections and 44,000 deaths annually, with the majority of outbreaks in developing countries linked to water contamination. Environmental monitoring, particularly wastewater surveillance, plays a key role in detecting HEV outbreaks. This study outlines a methodology for molecular detection of HEV in water samples, employing ammonium sulfate precipitation for virus concentration, magnetic bead-based nucleic acid extraction, and RT-nested PCR for high-sensitivity detection.

Keywords HEV, Water, Environmental monitoring, Virology, Molecular biology, PCR

1 Introduction

Hepatitis E virus (HEV) (family *Hepeviridae*; genus *Paslahepevirus*; species *Paslahepevirus balayani*) is responsible for self-limiting acute hepatitis in humans and is recognized as an emerging zoonotic disease [1, 2]. Virions are icosahedral, quasi-enveloped or non-enveloped, spherical particles with a diameter of 27–34 nm; and viral genomes are positive-sense monopartite RNA of 6.4–7.2 kb containing three ORFs [2]. It was previously referred to as the enterically transmitted non-A, non-B hepatitis virus, and it was first identified in 1983 [3, 4]. *Paslahepevirus balayani* has been assigned into eight distinct genotypes, labeled HEV-1 through HEV-8 [5]. Among them, HEV-1, HEV-2, HEV-3, and HEV-4 are the genotypes most frequently linked to HEV infections in humans. While HEV-1 and HEV-2 are specific to humans, HEV-3 and HEV-4 have a wider host range and are zoonotic. HEV is transmitted by contaminated water, the consumption of undercooked or raw meat or iatrogenic ally through blood transfusion or organ transplantation. Globally, HEV is a major cause of acute hepatitis, and the infection may become chronic in

Marciane Magnani and Geany Targino de Souza Pedrosa (eds.), *Detection and Quantification of Bacteria, Yeast, Viruses, and Protozoan in Foods and Freshwater*, Methods and Protocols in Food Science, https://doi.org/10.1007/978-1-0716-5064-6_12,

immunocompromised individuals and extrahepatic manifestations have been described [2]. Approximately 20 million HEV infections occur globally annually, leading to around 3.3 million symptomatic cases. In 2015, HEV accounted for approximately 44,000 deaths, representing 3.3% of the mortality associated with viral hepatitis [6, 7]. In many developing countries, contaminated drinking water is a primary cause of HEV outbreaks, predominantly involving genotypes 1 and 2 [8]. Large waterborne outbreaks have been reported in Africa, Central America, South and Central Asia, and tropical East Asia [7]. HEV-3 and 4 are primarily transmitted zoonotically through undercooked meat from pigs, wild boar, and deer [9]. In a recent wastewater surveillance study, HEV-3c was identified as the most prevalent variant [10]. The detection of viral genomes in wastewater serves as a valuable tool for assessing outbreak situations and conducting epidemiological surveillance, including monitoring viral variants [10].

Given the importance of monitoring the Hepatitis E virus in water, this Chapter will outline a methodology for its molecular detection from water samples. For virus concentration, we selected the ammonium sulfate precipitation method due to its simplicity, short processing time, and the availability of ammonium sulfate in large volumes. Additionally, this method is effective for virus recovery, making it a reliable choice [11]. For nucleic acid extraction and purification, we chose the magnetic beads method. Magnetic beads provide exceptional binding capacity, improved elution in small volumes, broad sample range compatibility, and versatility for both low and high-throughput applications, making these important advantages when compared to other DNA/RNA extraction and purification methods. We chose RT-nested PCR due to its high sensitivity, making it suitable for detecting viruses in environmental water samples [12]. Additionally, for the Nested-PCR we selected two sets of degenerate oligonucleotides designed to amplify HEV ORF1 because they are sensitive against several HEV strains [13] since members of different species in the genus are phylogenetically distinct based upon analysis of ORF1 codon positions 1–450 (methyltransferase), ORF1 codon positions 971–1692 (RNA-directed RNA polymerase) and ORF2 codon positions 121–473 (capsid protein) [2].

2 Materials

2.1 Viral Concentration

1. Ammonium sulfate.
2. TE buffer (Tris-EDTA).
3. Centrifuge bottles.
4. Micropipettes.
5. DNAse-free RNAse-free barrier tips.

6. microtube (for resuspension and storage).
7. Orbital shaker.
8. Centrifuge (capable of 10,000 ×*g*, with temperature control at 4°C).

2.2 Nucleic Acid Extraction

1. 1.5 mL or 2.0 mL DNAse-free RNAse microtubes.
2. DNAse-free RNAse-free barrier tips: 10 μL, 20 μL, 100 μL, 200 μL, and 1 mL (in racks).
3. King Fisher Deep Well extraction plates for 24 samples.
4. King Fisher Duo 12 Tip Comb.
5. MagMAX™ CORE Nucleic Acid Purification Kit (Applied Biosystems, Ref: A32700).
6. Vortex mixer.
7. Minicentrifuge (for quick spins).
8. Laminar flow.
9. Sample storage refrigerator.
10. King Fisher Duo Prime automated extraction system (Thermo Scientific™).

2.3 cDNA Synthesis

1. GoScript™ Reverse Transcription Mix, Random Primers (Promega Corporation, USA).
2. Nuclease-free, low retention qPCR-compatible reaction tubes or plates.
3. Pipettes and sterile aerosol-resistant tips.
4. Experimental and reference RNA samples (previously extracted RNA).
5. Thermal cycler or 25, 42, and 70°C controlled-temperature heat blocks.
6. Ice or 4°C cooling blocks.
7. Vortex mixer.
8. Minicentrifuge.
9. Laminar flow.
10. Sample storage refrigerator.

2.4 Nested PCR

1. GoTaq® Colorless Master Mix.
2. Primers (Table 1).
3. Nuclease-free water.
4. Thermal cycler (PCR machine).
5. Micropipettes.
6. Tips.

Table 1
Oligonucleotide used to perform Nested-PCR

Oligonucleotides	Sequence	Amplicon length (bp)
HEVORF1con-s1	5′-CTG GCA TYA CY CTA CTG CYA TTG AGC-3′	418
HEVORF1con-a1	5′-CCA TCR ARR CAG TAA GTG CGG TC-3′	
HEVORF1con-s2	5′-CTG CCY TKG CGA ATG CTG TGG-3′	287
HEVORF1con-a2	5′-GGC AGW RTA CCA RCG CTG AAC ATC-3′	

7. PCR tubes 0.2 mL.
8. PCR cabinet.
9. Vortex mixer.
10. Microcentrifuge tube spinner.

2.5 Electrophoresis

1. Agarose.
2. 1× TBE buffer.
3. Ethidium bromide.
4. DNA ladder.
5. Loading dye.
6. Electrophoresis system.
7. UV transilluminator.
8. Micropipettes.
9. Tips.
10. Flask/Beaker.

3 Methods

3.1 Viral Concentration Procedure

1. Add 60 ± 1 g of ammonium sulfate to a centrifuge bottle containing 150 mL of the water sample.
2. Dissolve the ammonium sulfate by gently inverting the bottles several times or by shaking them on an orbital shaker at 200 rpm until fully dissolved.
3. Incubate the samples at 3 ± 2°C for 60–180 min.
4. Before centrifugation, mark the outer sides of the bottle to identify where the pellets are expected to form.

5. Centrifuge the chilled samples for 30 min at 10,000 × g and 4 ± 1°C with no breaking to avoid disturbing the viral pellet.
6. Carefully remove the bottles from the centrifuge, ensuring the viral pellets are not disturbed (consider that on some occasions pellets might not be visible).
7. Decant the supernatant into a biohazard waste container.
8. Add 2 mL of TE buffer (Tris-EDTA) directly to the pellet and resuspend thoroughly by repeated pipetting.
9. Transfer the resuspended pellet into clean microtubes for subsequent steps.
10. If nucleic acid extraction is not to be performed immediately, store the sample at −80°C.

3.2 Nucleic Acid Extraction Procedure

Viral RNA extraction is performed by a commercial kit MagMAX-™CORE Nucleic Acid Purification Kit (Applied biosystems™, Thermo Fisher Scientific, Waltham, MA, USA), using automated equipment KingFisher™ Duo Prime (Thermo Fisher Scientific™), following the manufacturer's protocol. (https://assets.thermofisher.com/TFS-Assets/LSG/manuals/MAN0015944_MagMAXCORE_NA_Kit_UG.pdf)

3.2.1 Set Up the Processing Plates

1. Plate setup: KingFisher™ Flex instrument.
2. (Optional) To prevent evaporation and contamination, cover the prepared processing plates with sealing foil until they are loaded into the instrument.

Plate ID	Plate position[a]	Plate type	Reagent	Volume per well
Wash Plate 1	2	Deep Well	MagMAX™ CORE Wash Solution 1	500 µL
Wash Plate	3	Deep Well	MagMAX™ CORE Wash Solution 2	500 µL
Elution	4	Standard	MagMAX™ CORE Elution Buffer	90 µL
Tip Comb	5	Standard	Place a tip comb on the plate.	

[a]Position on the instrument

3.2.2 Prepare Bead/PK Mix

1. We recommend that you prepare a new Bead/PK Mix for each processing run. If necessary, you can store Bead/PK Mix at 4°C for up to 1 week.
2. Vortex the MagMAX™ CORE Magnetic Beads thoroughly to ensure that the beads are fully resuspended.

3. Combine the following components for the required number of samples, plus 10% overage (recommended).

Component	Volume per sample
MagMAX™ CORE Magnetic Beads	20 μL
MagMAX™ CORE Proteinase K	10 μL
Total Bead/PK Mix	30 μL

3.2.3 Prepare Lysis/Binding Solution

1. Combine the following components for the required number of samples plus 10% overage.

Component	Volume per sample
MagMAX™ CORE Lysis Solution	350 μL
MagMAX™ CORE Binding Solution	350 μL
Total Lysis/Binding Solution	700 μL

2. Mix by inverting the tube or bottle at least 10×.
3. (Optional) Store Lysis/Binding Solution at room temperature for up to 24 h.

3.2.4 Prepare the Sample

1. Prepare samples according to the sample type described above.

3.2.5 Combine the Sample with Bead/PK Mix and Lysis/Binding Solution

1. Invert the tube of Bead/PK Mix several times to resuspend the beads, then add 30 μL of the Bead/PK Mix to the required wells in the plate or tube strip.
2. Transfer the appropriate volume of each prepared sample (200 μL) to a well with Bead/PK Mix.
3. Mix the sample with the Bead/PK Mix for 2 min at room temperature according to your mixing method.
4. Add 700 μL of Lysis/Binding Solution to each sample-containing well or tube.
5. Immediately proceed to load the samples into the instrument.

3.2.6 Process Samples on the Instrument

1. Select the appropriate script on the instrument (see "Download and install the script" on MagMAX™ CORE Nucleic Acid Purification Kit User Guide).
2. Start the run, then load the prepared plates or tube strips in the appropriate positions when prompted by the instrument.
3. Store purified nucleic acid on ice for immediate use, or at −80°C for long-term storage.

3.3 cDNA Synthesis Procedure

After RNA extraction, Reverse transcription is conducted using the GoScript™ Reverse Transcription Mix, Random Primers (Promega Corporation, USA) following the manufacturer's instructions (https://www.promega.com.br/-/media/files/resources/protocols/product-information-sheets/a/9pia280-go-script-reverse-transcription-mix_random-primers.pdf?rev=7eb3c166c37c41a38d8e915f50a4111d&sc_lang=en).

3.3.1 First-Strand cDNA Synthesis

1. Thaw the GoScript™ Reverse Transcription Mix components on ice, mix gently and centrifuge briefly.
2. Prepare 10 μL of GoScript™ Reverse Transcription Mix for each reaction on ice, combining the components in the following order. Mix gently after each addition:

Component	Volume
Nuclease-free water	4 μL
GoScript™ Reaction Buffer, Random Primer	4 μL
GoScript™ Enzyme Mix	2 μL
Final volume	10 μL

3. Gently mix the GoScript™ Reverse Transcription Mix by pipetting or vortexing.
4. Optional: To denature RNA with high degrees of secondary structure (>60% GC content), incubate RNA samples at 70°C for 5 min, then immediately chill in ice water for at least 5 min. Centrifuge in a microcentrifuge for 5 s.
5. Combine 10 μL of GoScript™ Reverse Transcription Mix and up to 10 μL of RNA sample in each reaction tube or well of a reaction plate on ice as shown below.

Component	Volume
GoScript™ Reverse Transcription Mix	10 μL
Experimental or Reference RNA	10 μL[a]
Nuclease-free water	To a final volume of 20 μL[b]
Final volume	20 μL

[a]If using a sample volume of less than 10 μL, adjust the volume by adding nuclease-free water to reach a final reaction volume of 20 μL

[b]You will only use water if the sample volume is less than 10 μL

6. Mix each reverse transcription reaction by pipetting. Close or seal the reaction tubes or wells.
7. Incubate the reverse transcription reactions in controlled-temperature blocks or in a programmed thermal cycler.

Step	Temperature	Time	Number of cycles
Anneal primer[a]	25°C	5 min	1 cycle
Extension	42°C[b]	60 min	1 cycle
Inactivation	70°C[c]	15 min	1 cycle
Hold	4°C	∞	1 cycle

[a]Optional

[b]The extension temperature may be optimized between 37 and 55°C

[c]Higher temperatures up to 95°C for 5 min may be used for the inactivation step

8. Store the reaction products at 4°C in the reaction tubes or wells for immediate analysis (up to 24 h) or at −20°C for long-term storage.

3.4 Nested PCR Procedure

1. Prepare the first master mix by combining the GoTaq® Colorless Master Mix, 2×, primers, and nuclease-free water in a PCR tube, keeping it on ice.
2. For a final reaction volume of 50 μL, use the following proportions:

Components	Volume per reaction (μL)
GoTaq® Colorless Master Mix, 2×	25
Forward primer (20 μM stock)[a]	1
Reverse primer (20 μM stock)[b]	1
Nuclease-free water	18

[a]HEVORF1con-s1

[b]HEVORF1con-a1

3. Add 5 μL of DNA template (cDNA product) to the reaction mix.
4. Mix the reaction gently and spin briefly to bring the contents to the bottom of the tube.
5. Place the reaction tube into the thermal cycler and start the PCR program:

Step	Temperature (°C)	Time	Cycles
Denaturation	95	5 min	1
Denaturation, annealing, and extension	95	30 s	45
	59	1 min	
	72	1 min	
Final extension	72	7 min	1
Hold	4	Hold	∞

3.5 Electrophoresis

1. In a Flask/Beaker add 2 g of agarose powder and 100 mL of 1× TBE buffer.
2. Heat the mixture in a microwave in short bursts (~20–30 s) until the agarose is completely dissolved.
3. For each mL of TBE buffer use 0.1 μL of ethidium bromide, in this case to 100 μL used 10 μL.
4. Pour the agarose solution into the gel casting tray and insert the comb to create wells. At room temperature wait until the gel solidifies.
5. Place the solidified gel into the electrophoresis tank and fill the tank with 1× TBE buffer until the gel is submerged.
6. Carefully load the DNA ladder (molecular weight marker) by mixing it with loading dye as recommended by the manufacturer.
7. Carefully load the 10 μL of the mixture of 10 μL the PCR products with 2 μL of 6× loading dye.
8. Connect the gel apparatus to the power supply and run the gel at 30 volts for 10 min, following 70 volts for 60 min.
9. After the run is complete, remove the gel and view it using a UV transilluminator to visualize the DNA bands.
10. Compare the PCR product bands with the positive control and the DNA ladder to determine the size of the amplified products.

4 Notes

4.1 Viral Concentration

1. If the ammonium sulfate is not fully dissolved, viral precipitation may be inefficient, leading to poor recovery. Ensure thorough mixing by inverting the bottle or shaking on an orbital shaker until the solution is completely clear. If necessary, increase shaking time slightly but avoid excessive agitation.
2. Mark the bottles where the pellet is expected to form before centrifugation, because the Viral pellets are often invisible, and the pellet may be disturbed or lost during the supernatant removal step.
3. Variations in incubation or centrifugation temperatures can impact on the efficiency of viral precipitation and concentration. Pre-chilling the centrifuge rotor to ensure consistent temperature throughout the run.

4.2 Nucleic Acid Extraction

1. Ensure that all steps are performed with DNAse and RNAse-free materials.

2. Check your MagMAX™ CORE kit to ensure all components (lysis buffer, wash buffers, elution buffer, magnetic beads, etc.) are available before starting the extraction process.

4.3 Nested PCR

1. Primer dimers or non-specific products may appear. Optimize primer concentrations and annealing temperature, consider using a gradient PCR to find the best annealing temperature.
2. Nested PCR is sensitive to contamination, which can lead to false positives. Always use negative control and handle samples carefully to avoid contamination.
3. It's possible that the sample contains inhibitors that affect the PCR reactions. Performing serial dilutions of the sample or cDNA may significantly reduce this inhibition and improve amplification efficiency.
4. Additional sensitivity can be achieved by employing a touch-down PCR approach, which optimizes amplification by gradually decreasing the annealing temperature during the initial cycles.

4.4 Electrophoresis

1. Ethidium bromide is a hazardous chemical; handle it with care, using appropriate PPE.

4.5 Positives Samples

1. Since HEV is divided into multiple genotypes, sequencing and characterization of positive samples are crucial for identifying the specific HEV strains in circulation.

Acknowledgments

Our studies are funded by grants from Fapergs (no 23/2551-0002221-4) and CNPq (no 405786/2022-0).

References

1. Pischke S, Stiefel P, Franz B, Bremer B, Suneetha PV, Heim A, Ganzenmueller T, Schlue J, Horn-Wichmann R, Darnedde M, Scheibner Y, Raupach R, Scheibner Y, Taubert R, Haverich A, Manns MP, Wedemeyer H, Bara CL (2012) Chronic hepatitis E in heart transplant recipients. Am J Transplant 12:3128–3133
2. Purdy MA, Harrison TJ, Jameel S, Meng XJ, Okamoto H, Van der Poel WHM (2022) ICTV Report Consortium. ICTV virus taxonomy profile: Hepeviridae. J Gen Virol 98:2645–2646
3. Balayan MS, Andjaparidze AG, Savinskaya SS, Ketiladze ES, Braginsky DM, Savinov AP, Poleschuk VF (1983) Evidence for a virus in non-A, non-B hepatitis transmitted via the fecal-oral route. Intervirology 20:23–31
4. Mirazo S, Ramos N, Arbiza J (2012) Molecular epidemiology of hepatitis E virus (HEV) in South America: current status. Virus Rev Res 17:1
5. Smith DB, Izopet J, Nicot F, Simmonds P, Jameel S, Meng XJ, Norder H, Okamoto H, van der Poel W, Reuter G, Purdy MA (2020) Update: proposed reference sequences for subtypes of hepatitis E virus (species Orthohepevirus A). J Gen Virol 101:692–698
6. World Health Organization (2023) Hepatitis E. https://www.who.int/news-room/fact-sheets/detail/hepatitis-e (accessed 30 Aug 2024)
7. Teshale E (2024) Hepatitis E. CDC Yellow Book 2024: Travel-Associated Infections &

Diseases. https://wwwnc.cdc.gov/travel/yellowbook/2024/infections-diseases/hepatitis-e#:~:text=Epidemiology,Asia%2C%20and%20tropical%20East%20Asia (accessed 30 Aug 2024)
8. Hakim MS, Wang W, Bramer WM, Geng J, Huang F, de Man RA, Peppelenbosch MP, Pan Q (2017) The global burden of hepatitis E outbreaks: a systematic review. Liver Int 37:19–31
9. Ricci A, Allende A, Bolton D, Chemaly M, Davies R, Fernandez Escamez PS, Herman L, Koutsoumanis K, Lindqvist R, Nørrung B, Robertson L, Ru G, Sanaa M, Simmons M, Skandamis P, Snary E, Speybroeck N, Ter Kuile B, Threlfall J, Wahlström H, Di Bartolo I, Johne R, Pavio N, Rutjes S, van der Poel W, Vasickova P, Hempen M, Messens W, Rizzi V, Latronico F, Girones R (2017) Public health risks associated with hepatitis E virus (HEV) as a food-borne pathogen. EFSA J 15:e04886
10. Rau F, Elsner C, Meister TL, Gömer A, Kallies R, Dittmer U, Steinmann E, Todt D (2024) Monitoring of hepatitis E virus in wastewater can identify clinically relevant variants. Liver Int 44:637–643
11. Kevill JL, Pellett C, Farkas K, Brown MR, Bassano I, Denise H, McDonald JE, Malham SK, Poter J, Warren J, Evens NP, Paterson S, Singer AC, Jones DL (2022) A comparison of precipitation and filtration-based SARS-CoV-2 recovery methods and the influence of temperature, turbidity, and surfactant load in urban wastewater. Sci Total Environ 808:151916
12. Kittigul L, Ekchaloemkiet S, Utrarachkij F, Siripanichgon K, Sujirarat D, Pungchitton S, Boonthum A (2005) An efficient virus concentration method and RT-nested PCR for detection of rotaviruses in environmental water samples. J Virol Methods 124:117–122
13. Erker JC, Desai SM, Mushahwar IK (1999) Rapid detection of Hepatitis E virus RNA by reverse transcription-polymerase chain reaction using universal oligonucleotide primers. J Virol Methods 81:109–113

Chapter 13

Molecular Detection of *Salmonella* in Meat Products

Débora Rodrigues Silveira and Graciela Volz Lopes

Abstract

Salmonellosis is one of the major foodborne diseases. Particularly *Salmonella enterica* and its serovars Enteritidis and Typhimurium, are significant pathogens causing nontyphoidal salmonellosis in humans, often transmitted through contaminated food products. Timely detection of these pathogens is crucial for food safety and public health. This chapter describes a multiplex PCR (mPCR) method for the rapid detection of *Salmonella* Enteritidis and Typhimurium in meat products. The method involves pre-enrichment of samples in buffered peptone water, DNA extraction using the phenol-chloroform method, and PCR amplification employing a primer set targeting the *srfC* gene specific to *Salmonella enterica*, along with two additional primer sets, STM4495 and SEN1392, targeting *S.* Typhimurium and *S.* Enteritidis, respectively. The mPCR reaction is optimized to detect as low as 23 CFU/PCR for *S.* Typhimurium and 22 CFU/PCR for *S.* Enteritidis, offering a rapid and reliable diagnostic tool for foodborne pathogen surveillance. The effectiveness of this method is demonstrated through gel electrophoresis, producing distinct DNA fragments for each target. This approach provides a sensitive, specific, and efficient alternative to traditional culture-based detection methods, supporting enhanced food safety monitoring and public health initiatives.

Keywords Multiplex PCR, Typhimurium, Enteritidis, Molecular biology, Pathogen

1 Introduction

Salmonella is a Gram-negative, rod-shaped, facultative anaerobic genus, belonging to the Enterobacteriaceae family and are comprised of two species, *Salmonella enterica* and *Salmonella bongori*, conforming to the White-Kauffmann scheme [1]. *Salmonella enterica* subsp. *enterica* is the most relevant in terms of carriage in animals used for food production, while the other subspecies and *S. bongori* are found in the environment and cold-blooded animals [2]. The species *S. enterica* is comprised of six subspecies, *enterica, salamae, arizonae, diarizonae*, *houtenae*, and *indica*, with more than 2600 serovars. Of these, the *enterica* subspecies is responsible for approximately 1550 serovars, of which 99% may cause infections in both animals and humans [3].

Marciane Magnani and Geany Targino de Souza Pedrosa (eds.), *Detection and Quantification of Bacteria, Yeast, Viruses, and Protozoan in Foods and Freshwater*, Methods and Protocols in Food Science, https://doi.org/10.1007/978-1-0716-5064-6_13,

The most common human nontyphoidal salmonellosis is caused by *S. enterica* serovar Enteritidis followed by Typhimurium, Newport, Heidelberg, and Montevideo. The food vehicles associated with this serovars include a wide variety of products including eggs, chicken, pork, leafy greens, peanut butter, turkey, dairy products, and vegetables [4]. An estimated 1.35 million human cases of salmonellosis occur annually in the United States every year [5]. The most common human nontyphoidal salmonellosis in the US is caused by *S. enterica* serovars Enteritidis and Typhimurium [6]. Consequently, it is important to isolate, detect and identify these serovars from contaminated foodstuffs to ensure food safety.

To detect *Salmonella* in food samples in general, molecular testing in addition to conventional culture-based methods can be used [7]. Multiplex PCR (mPCR) is a variation of reaction chain polymerase, which involves more than one pair of primers allowing one single reaction to detect more than one type of microorganism along with their serotypes and/or different genes, therefore reducing the diagnosis time [8]. Considering the importance of this group of bacteria in the processing of meat, its importance to public health, and considering the need of fast and accurate diagnosis methods, we described a multiplex PCR method to detect genus *Salmonella* and serovars Enteritidis and Typhimurium in meat products.

2 Materials

2.1 Pre-Enrichment Media

Different media have been reported successful for the simultaneous recovery of *Salmonella* spp. (*see* **Note 1**). Buffered peptone water (BPW) and lactose broth (LB) are commonly used, but others media, such as soy tryptone and nutrient broth, can be employed. A list of some of these media and incubation conditions is provided in Table 1.

For the present application BPW was selected, as it has been reported to be suitable for many different food types. The composition of this broth is (g/litre: 10.0 peptone (includes enzymatic digest of casein), 5 g Sodium chloride, 1.5 g Potassium dihydrogen phosphate; 9.0 g di-Sodium hydrogen phosphate dodecahydrate. Final pH 7.0 ± 0.2 at 25°C.

Table 1
Culture media

Pre-enrichment media	Incubation condition	Method
Buffered peptone water (BPW)	34°C to 38°C for 18 h	ISO 6579-1 [9]
Lactose broth (LB)	35 ± 2°C for 24 ± 2 h	BAM/FDA [10]
Lactose broth (LB)	35 ± 2°C for 18 ± 24 h	APHA [11]

ISO, International Organization for Standardization; *BAM/FDA*, Bacteriological Analytical Manual da Food and Drug Administration; *APHA*, American Public Health Association

2.2 Chemicals

1. TE buffer 10×: 100 mM Tris-Cl (desired pH), 10 mM EDTA (pH 8.0). Sterilize solutions by autoclaving for 20 min at 15 psi (1.05 kg/cm^2) on liquid cycle. Store the buffer at room temperature.
2. Sodium n-Dodecyl Sulfate solution (SDS) (10%, w/v).
3. Proteinase K (20 mg/mL in TE, pH 7.5).
4. Phenol:chloroform (1:1, v/v).
5. Chloroform.
6. Chloroform/Isoamyl alcohol 24:1.
7. NaCl 5 M.
8. Ethanol (70% and 95%).
9. Ribonuclease A (5 mg/mL in TE, pH 8.0).
10. Ammonium acetate 5 M.
11. Isopropanol.
12. Primers (Table 2).
13. Master Mix components for multiplex PCR reaction (Table 3).
14. Sterile Molecular Grade water.
15. DNA staining solution.
16. 6× Gel-loading buffer.

2.3 Equipments and Materials

1. PCR thermocycler with gradient thermal cycler system. In the described methodology a PTC-200 thermal cycler (BIO-RAD, USA) was used.
2. Centrifuge with a speed range of 380–16,000 ×*g*. Also, must have a rotor suitable for 1.5–2 mL tubes.

Table 2
Specific primer sets of multiplex PCR for identifying *Salmonella enterica* Typhimurium and Enteritidis

Target	Gene	Primer	Sequence 5′–3′	Size (bp)	Tm (°C)
S. Typhimurium	STM4495	stm-4495	F: GGTGGCAAGGGAATGAA R: CGCAGCGTAAAGCAACT	915	65
S. Enteritidis	SEN1392	sen-1392	F: GCCACTGTCTGATGCTCTTG R: GAAAGGCTCCGTGGTTAGT	656	60
S. enterica	*srfC*	FS23	F: GGCGTTAACCCACTCCAGTA R: TTACTGTGGGGAGAGCAACC	492	58

F, primer forward; *R*, primer reverse; *bp*, base pair

Table 3
Master mix for multiplex PCR.

Reagent	Final concentration	Volume (μL)
Buffer 10× (free Mg^{2+})	1×	3.0
$MgCl_2$	2.5 mM	1.5
dNTPs	2.5 mM	2.5
Primer STM4495 F	5 μM	1.0
Primer STM4495 R	5 μM	1.0
Primer SEN1392 F	5 μM	1.0
Primer SEN1392 R	5 μM	1.0
Primer *srfC* F	5 μM	1.0
Primer *srfC* R	5 μM	1.0
Taq DNA Polymerase	1 U/μL	1.0
DNA template	5 ng	2.5
Ultrapure H_2O	–	14.5
Total		30

F, primer forward; *R*, primer reverse

3. Stomacher® or other type of sample homogenizer (e.g., Smasher®), with compatible filter bags.
4. Thermomixer (dry bath) capable of shaking and heating.
5. Scale suitable for a weight range of 25 g to 2 Kg.
6. Vortex Mixer.
7. Pipets (1–1000 μL volume).
8. Pippete tips (0.2–1000 μL volume) (aerosol resistant tips).
9. Microcentrifuge tubes (0.2–0.5 mL).
10. Microcentrifuge tubes (1.5 mL).
11. Spectrophotometer.
12. Cuvettes for measuring absorbance at 260 nm.

3 Methods

3.1 Sample Pre-Enrichment

1. Aseptically weigh 25 g sample into sterile bags.
2. Add 9× mL of diluent (i.e., 225 mL of BPW).
3. Homogenize the matrix for 30–120 s in a laboratory homogenizer (i.e., Stomacher).

4. Incubate the mixture at 35 ± 1°C for a minimum of 18 h (no deleterious effects on the performance of the method have been observed in the incubation time is extended).
5. After incubation, take an aliquot (1–2 mL) for DNA extraction.

3.2 DNA Extraction

The isolation of DNA from bacteria based on the Phenol-Chloroform method was described by Green and Sambrook [12].

1. Transfer 1.5 mL of an overnight bacterial culture to a 1.5 mL microcentrifuge tube. Centrifuge the filled tube for 30 s at room temperature. Remove the supernatant by aspiration.
2. Add 400 μL of TE (pH 8.0). Disperse the bacterial pellet by vortexing twice for 20 s each time.
3. Add 50 μL of 10% SDS, 50 μL of proteinase K (20 mg/mL in TE, pH 7.5). Incubate the bacterial lysate for 1 h at 37°C.
4. At this stage, the digested lysate is too viscous to handle. Shear the DNA by three to five passages through a 26-gauge needle. Try to minimize foaming.
5. Add 500 μL of a 1:1 mixture of phenol:chloroform (*see* **Notes 2 and 3**).
6. Close the top of the microcentrifuge tube, and mix the two phases by gentle vortexing. At the interface, a more or less pronounced and light-colored protein layer is visible (*see* **Note 4**).
7. Transfer the upper layer to a fresh microcentrifuge tube, and repeat Steps 5 and 6 two or three times, until no material is visible at the interface.
8. Extract the aqueous phase twice with 500 μL of chloroform.
9. Add equal volume (c.a. 500 μL) of Chloroform/Isoamyl alcohol 24:1. Mix vigorously for 2 min.
10. Transfer the supernatant to a fresh microcentrifuge tube and add ~25 μL of 5 M NaCl and 1 mL of 95% ethanol. Vortex and then centrifuge for 10 min at 4°C.
11. Dissolve the damp-dry pellet in 100 μL of TE (pH 8.0), add 5 μL of ribonuclease A (5 mg/mL in TE, pH 8.0). Incubate the solution for 30 min at 37°C.
12. Add 40 μL of 5 M ammonium acetate and 250 μL of isopropanol. Vortex briefly and store the closed tube for 10 min at room temperature.
13. Recover the DNA by centrifugation (10 min at room temperature). Wash the pellet twice with 70% ethanol. Decant the supernatant, and then carefully remove the residual ethanol with a micropipette. Leave the open tube on the bench for ~15 min.

14. Dissolve the damp-dry pellet in 100 μL of TE (pH 8.0). Measure the concentration of DNA from its absorbance of a 1:10 dilution of a sample of the DNA (in H_2O) at 260 nm. As a blank, use a 1:10 dilution of TE (pH 8.0) in H_2O: 1OD260 = 50 mg/mL of DNA.
15. Store the DNA, depending on its intended use (*see* **Note 5**).

3.3 Multiplex PCR

1. Three primer sets are recommended in this multiplex PCR: the primer set stm-4495 for *S.* Typhimurium, the primer set sen1392 for *S.* Enteritidis, and the primer set FS23 specific for *S. enterica*, used to discriminate from non-*Salmonella* strains. These primers were designed by Liu et al. [13]. The oligonucleotide sequences are provided in Table 2.
2. Prepare the reaction mixture according to Table 3 (*see* **Note 6**). Prepare enough master mix to analyze every sample, at least, in duplicate (technical duplicates), as well as a positive and a negative control. It is advisable to always prepare some excess, i.e., 1 extra reaction.
3. A PTC-200 thermal cycler (BIO-RAD, USA) was used for the present method, but other thermocyclers are also available in the market. Some thermocyclers may require a mineral oil overlay (50–70 μL). The thermocycler conditions are provided in Table 4.
4. The multiplex PCR is capable of detecting *Salmonella* cells as few as 23 CFU/PCR for *S.* Typhimurium or 22 CFU/PCR for *S.* Enteritidis [12].
5. An aliquot (5 μL) from the PCR products of each sample should be mixed with 2 μL 6× gel loading buffer and analyzed by gel electrophoresis in 2% agarose gels.

Table 4
The following thermocycler conditions should be used:

Thermocycler conditions	Time	Temperature (°C)	Number of cycles
Initial denaturation	5 min	94	1
Denaturation	30 s	94	10 (−0.5°C per cycle from 65 to 60.5°C)
Primer annealing	30 s	65	
Primer extension	30 s	72	
Denaturation	30 s	94	25
Primer annealing	30 s	60	
Primer extension	30 s	72	
Final extension	10 min	72	1

6. Run the gel under the following electrophoresis conditions: 80 V; 150 mA; 150 W; 70 min.
7. Stain the gel by immersing in the electrophoresis buffer or H_2O containing ethidium bromide (0.5 μg/mL) for 30–45 min at room temperature.
8. Illuminate gel with a UV transilluminator and visualize bands relative to molecular weight marker migration. Polaroid photographs can be taken of the gel for documentation.
9. The primers listed give a 915 bp fragment for *S.* Typhimurium, 656 bp fragment for *S.* Enteritidis, and 492 bp fragment for *S. enterica*.
10. Positive and negative culture controls and reagent control should be included with each PCR run.

4 Notes

1. *Salmonella* can be present in low numbers and is often accompanied by considerably larger numbers of other Enterobacteriaceae or bacteria of other families. Pre-enrichment is used to permit the detection of low numbers of *Salmonella* or injured *Salmonella*.
2. Phenolic solutions should be completely clear and non-colored and can be stored for approx. 1 year in a refrigerator (dark and under steady temperature). In case a more or less pronounced reddish or pink stain is visible, the phenol has started oxidizing and is likely to destroy the nucleic acids during the purification process. Discard and replace by a new batch!
3. Phenol should have a pH of 7.5–8.0 for DNA isolation.
4. Normally, the aqueous phase is on top. However, special composition of the aqueous phase may lead to inversion of the phases (see above). The organic phase is easily identifiable because of the yellow color resulting from the hydroxyquinoline added to the phenol.
5. DNA can be stored indefinitely at −20°C or at 4°C in TE buffer (pH 8). There is no difference in stability between DNA stored in TE at −20°C or at −80°C. However, DNA stored in pure water is susceptible to acid hydrolysis. Tubes used for long-term storage of DNA should be tightly sealed to prevent evaporation.
6. To minimize cross-contamination of PCR reagents, it is recommended that a PCR master mix be prepared and aliquots stored frozen (−20°C) until use. Master mixes contain all necessary reagents, except Taq polymerase and the lysate (template) to be amplified.

Acknowledgments

This work was supported by Coordenação de Aperfeiçoamento de Pessoal de Nível Superior – CAPES, Conselho Nacional de Desenvolvimento Científico e Tecnológico – CNPQ, and Fundação de Amparo a Pesquisa do Estado do Rio Grande do Sul – FAPERGS.

References

1. Popoff MY, Bockemühl J, Gheesling LL (2004) Supplement 2002 (no. 46) to the Kauffmann–White scheme. Res Microbiol 155:568–570
2. Eng SK, Pusparajah P, Ab Mutalib NS, Ser HL, Chan KG, Lee LH (2015) *Salmonella*: a review on pathogenesis, epidemiology and antibiotic resistance. Front Life Sci 8:284–293
3. Cavestri C, Savard P, Fliss I, Emond-Rhéault J, Hamel J, Kukavica-Ibrulj I, Boyle B, Daigle F, Malo D, Bekal S, Harris L, Levesque R, Goodridge L, LaPointe G (2022) *Salmonella enterica* subsp. *enterica* virulence potential can be linked to higher survival within a dynamic *in vitro* human gastrointestinal model. Food Microbiol 101:103877
4. Andino A, Hanning I (2015) *Salmonella enterica*: survival, colonization, and virulence differences among serovars. Sci World J 2015:520179
5. CDC, Center for Disease Control and Prevention (2024) Salmonella Infection (salmonellosis), https://www.cdc.gov/salmonella/index.html (accessed: 15 nov. 2024)
6. Nazari Moghadam M, Rahimi E, Shakerian A, Momtaz H (2023) Prevalence of *Salmonella* Typhimurium and *Salmonella* Enteritidis isolated from poultry meat: virulence and antimicrobial-resistant genes. BMC Microbiol 23:168
7. Heymans R, Vila A, van Heerwaarden CAM, Jansen CCC, Castelijn GAA, van der Voort M, Biesta-Peters EG (2018) Rapid detection and differentiation of *Salmonella* species, *Salmonella* Typhimurium and *Salmonella* Enteritidis by multiplex quantitative PCR. PLoS One 13:e0206316
8. Malkawi HI, Gharaibeh R (2003) Multiplex PCR for the direct detection of *Salmonella enterica* from chicken, lamb and beef food products. J Basic Microbiol 43:328–336
9. ISO 6579-1 (2017) Microbiology of the food chain - horizontal method for the detection, enumeration and serotyping of *Salmonella* - Part 1: Detection of *Salmonella* spp. ISO, Geneva, Switzerland
10. Andrews WH, Wang H, Jacobson A, Ge B, Zhang G, and Hammack T (2024) Bacteriological analytical manual chapter 5: *Salmonella*. Food & Drug Administration, Washington DC
11. APHA (2001) Standard methods for the examination of water and wastewater, 20th ed. American Public Health Association, Washington DC
12. Green MR, Sambrook J (2017) Isolating DNA from gram-negative bacteria. Cold Spring Harb Protoc 2017(1). https://doi.org/10.1101/pdb.prot093369.
13. Liu B, Zhou X, Zhang L, Liu W, Dan X, Shi C, Shi X (2012) Development of a novel multiplex PCR assay for the identification of *Salmonella enterica* Typhimurium and Enteritidis. Food Control 27:87–93

Check for updates

Chapter 14

Double-Layer Agar Plaque Protocol for Propagation and Quantification of SARS-CoV-2 Surrogate, Bacteriophage Phi6: Techniques and Applications

Ruthchelly Tavares da Silva, Edson Douglas Silva Pontes and Marciane Magnani

Abstract

Given biosafety concerns in virus research, the use of viral substitutes, such as bacteriophages, has been well accepted, presenting results comparable to those of the target viruses under the same experimental conditions. The double-layer agar plate (DLAP) protocol is based on the ability of bacteriophages to infect and lyse growing bacterial cells. In this method is necessary to use two layers of agar, where the first is the base layer and the second is a softer agar layer made of agar and a mix of host bacteria. DLAP protocol is a widely used, low-cost method for the enumeration and detection of bacteriophages in food, water, and beverages. In this chapter, we used the Phi6 SARS-CoV-2 surrogate as an example.

Keywords Bacteriophages, Phage enumeration, Host bacteria, Viral quantification, Plaque assay, Agar overlay technique

1 Introduction

The coronavirus disease caused by the SARS-CoV-2 virus has a high transmission capacity and lethality and was quickly considered a pandemic by the World Health Organization (WHO) [1]. Despite the decline of the pandemic, the fear of new diseases caused by viruses remains [2]. Although there is no evidence, to date, of the transmission of SARS-CoV-2 through water or food [3], research indicates the presence of SARS-CoV-2 genetic material on the surface of food and food packaging [4], as well as its ability to survive in water [5], food [6, 7], and packaging [8].

Despite the need, the development of studies with pathogenic viruses is extremely challenging, as it requires laboratories with a high level of biosafety, in addition to the potential health risks for

Marciane Magnani and Geany Targino de Souza Pedrosa (eds.), *Detection and Quantification of Bacteria, Yeast, Viruses, and Protozoan in Foods and Freshwater*, Methods and Protocols in Food Science, https://doi.org/10.1007/978-1-0716-5064-6_14,

researchers [9]. Therefore, the use of appropriate viral surrogates to reduce the complexity of the work has been a strategy adopted by many researchers [2]. The use of phages as viral surrogates has a number of advantages, such as low cost, short cultivation times, ease and safety, and they can be produced in large quantities [10, 11].

Double-layer agar plate (DLAP), widely used method for enumerating bacteriophages [12], as the name suggests, is a two-layer method. In this technique, the bottom layer (solid agar) is prepared with medium supporting bacterial growth, and the top layer (soft agar), contains the same medium with lower concentration of agar (0.4–0.6%) [13], mixed with the host bacteria and poured onto the bottom layer, resulting in the so-called lawn. Phages can be added to the second layer and dried or mixed directly with the host bacteria and soft agar [14]. After the incubation period (optimal time and temperature for bacterial growth), clear circular patches or plaques appear on the lawn, indicating that the phage has been able to infect the bacteria [15, 16]. The concentration of infectious phages (titer) in a phage stock or sample is expressed in plaque-forming units (PFU), where each plaque represents one phage, but in some cases a cluster of phages can produce a single plaque [17].

Enumeration of phages by the DLAP protocol is simple, efficient, applicable to many bacteriophages and does not require specific equipment or extensive training, making it easy to implement [14, 18]. However, it has some limitations, such as a long time to perform the analysis, a specific temperature control for soft agar is required, in addition to a high standardization of the method being necessary for good reproducibility [12, 16]. This chapter aims to describe the DLAP protocol for quantifying Phi6 SARS-CoV-2 surrogate.

2 Materials

1. Culturing tools: 10.0-, 1.0-, 0.1-, and 0.01-mL micropipettes (*see* **Note 1**), sterile pipette tips, sterile plastic 15–50-mL tubes, sterile and disposable inoculation loop, sterile microcentrifuge tube 1.5 mL, syringe filter (PVDF Hydrophilic Membrane, 0.22 μm) (*see* **Note 2**), syringes (sterile, disposable, 5, 10, or 20 mL), petri dishes (90 mm), and cell scraper.
2. Growth media: Luria-Bertani broth (LB) and bacteriological agar.
3. Reagents: sodium hydroxide solution (NaOH; 1 N)
4. Equipment: heating block or water bath, incubator/shaker capable of maintaining 25 ± 1.0°C for growth of microorganisms, centrifuge, and vortex mixer.

2.1 Luria-Bertani Broth

1. Prepare the LB medium according to the manufacturer's instructions.
2. Adjust pH of medium to 7.5 adding NaOH 1 N.
3. Heat sterilizes (*see* **Note 3**).
4. Fill sterile plastic 50-mL tubes with 10 mL liquid LB medium and store the tubes at 4 ± 1.0°C for up to 1 month.

2.2 Bottom Layer or Solid Agar Media

1. Prepare the LB medium according to the manufacturer's instructions, with addition of bacteriological agar to a concentration of 1.5% and heat sterilize (*see* **Note 3**).
2. After autoclaving, allow flasks to cool in a 55 ± 1.0°C water bath before pouring.
3. Dispense ~20 mL of medium per plate, and when cool, bag and store the plates at 4 ± 1.0°C for up to 1 month (*see* **Note 4**).

2.3 Top Layer or Soft Agar

1. Prepare the LB medium according to the manufacturer's instructions, with addition of bacteriological agar to a concentration of 0.7%.
2. Preheat the medium to melt the agar and mix the broth/agar.
3. Heat sterilizes the medium.
4. Dispense ~5 mL of medium into sterile plastic tubes (top agar), caps the tubes with plastic caps, and store at 4°C up to 1 month.
5. Before use, heat the tubes to 42 ± 1.0°C in a water bath or heating block (*see* **Note 5**).

2.4 Host Bacteria Working Stock Cultures

1. Remove the LB plaques from 4°C storage and dry (*see* **Note 6**).
2. Use sterile loop to remove a small amount of *P. syringae* pathovar *phaseolicola* (or equivalent bacteriophage host) from the frozen stock culture (*see* **Note** 7).
3. Immediately streak the inoculating loop very gently over the plate using a back-and-forth motion (*see* **Note 8**).
4. Incubate the plaques at 25 ± 1.0°C for two overnights.
5. Use sterile loop to transfer an isolated colony from quadrant streak plate to both in tube.
6. Place the tube into a sterile beaker (or flask) and incubate at 25 ± 1.0°C with shaking (100 rpm) overnight.
7. Harvest cells by centrifugation (5000 × *g*, 4°C for 25 min) and adjust the inoculum to the estimated level.
8. Keep 4 ± 1.0°C until use (*see* **Note 9**).

3 Methods

3.1 Double Agar Overlay Plaque Assay

1. Remove the LB plaques from 4°C storage and dry (*see* **Note 10**).
2. When the plaques are dry, number them sequentially with the dilutions to be plated (e.g., −1 to −11), and a control.
3. Turn water bath or heating block and allow time to equilibrate to temperature (42 ± 1.0°C).
4. Keep top agar tubes in the water bath or heating block and allow agar tubes to equilibrate to 42°C (*see* **Note 5**).
5. Add some bacterial culture (overnight culture at a ratio of 1:50 to top agar tube). Vortex briefly and pour the contents over the surface of LB plaques dried (*see* **Notes 10, 11, and 12**).
6. Wait until media has solidified (about 30 min).
7. Separate 11 sterile microcentrifuge tube and add 900 μL of LB broth (or equivalent diluent) in each tube and label them sequentially (e.g., −1 to −11).
8. Pipette 100 μL of Phi6 stock bacteriophage (or another bacteriophage) to the first tube. Vortex briefly. Change the pipette tip and transfer 100 μL to the second tube in the series (*see* **Note 13**).
9. Place plates in appropriated incubator (25 ± 1.0°C) overnight.
10. Count the plates on plates with 20–200 plates.

3.2 Harvesting a Lysate and Filter Sterilization (Using Syringe)

1. Pour 5 mL of LB broth onto plaques with visible regions of opaque bacterial lawn (*see* **Note 14**), scrape the top agar using a sterile cell scraper and place in a labeled 15-mL Falcon tube.
2. Vortex the Falcon tube for ~15 s to release the phage from the top agar.
3. Centrifuge at 3000 × *g*, 4°C for 10 min.
4. Sterilize liquid portion by filtration.
5. Screw the syringe barrel onto the 0.22-μm filter (*see* **Note 15**).
6. Place the syringe barrel + filter into a sterile plastic 5-mL tube.
7. Pour the supernatant from the Falcon tube into the syringe barrel and press the syringe plunger to pass the liquid through the filter.
8. Store the lysate in the refrigerator at 4°C.

3.3 Stocks

3.3.1 Pseudomonas Syringae Pathovar Phaseolicola Glycerol Stocks

1. Add 800 μL of *P. syringae* pathovar *phaseolicola* overnight culture to an autoclaved cryotube containing 200 μL of an 80% glycerol solution.
2. Stored at −20°C (*see* **Note 16**).

3.3.2 Phage Glycerol Stocks

1. Add 500 μL filtered phage lysate to an autoclaved cryotube containing 300 μL of an 80% glycerol solution and 200 μL of LB broth.
2. Stored at −20°C (*see* **Note 16**).

3.4 Data Analysis and Calculations

1. Count the number of plaque-forming units (PFU) in plaque counts within the desired range of 20–200 PFU.
2. Use the following equation for determine the titer of the original phage preparation:

4 Notes

1. Micropipettes must be cleaned before use and calibrated periodically.
2. We recommend using filters with PVDF membranes, but any 0.22-μm hydrophilic filter can be used.
3. Fill containers only halfway due to pressure changes during autoclaving, and briskly mixing the agar before autoclaving will facilitate dissolution.
4. To avoid condensation on the plates, the agar must be dried in the biological safety cabinet, with sterile air flowing directly above the agar.
5. Below 42°C, the top agar may solidify and affect plate counting ability.
6. LB plates should be dried in a appropriated incubator at 25°C/overnight and before use may be partially uncovered in a laminar flow hood for 10–15 min to reduce condensation prior to inoculation.
7. The laboratory should acquire reference host bacterial cultures and use them to establish pure, frozen, laboratory-maintained stock cultures.
8. The quadrant streak technique should be used to obtain well-isolated discrete colonies.
9. For best results use only culture stored at 4°C for no more than 24 h.
10. Mix the bacterial culture briefly (approximately 30 s is sufficient).
11. Each 5 mL of medium into sterile plastic tubes (top agar) must be poured on the surface of a LB plates.
12. The medium should be completely melted and free of lumps (swirl to obtain uniform coverage of the upper agar over the

lower agar). Avoid exposing the agar to room temperature too much to prevent premature solidification.

13. Vortex briefly the diluted phage tube for approximately 30 s. Long mixing times is not recommended for phage suspensions.
14. These plates generally produce the highest concentration of phage in a lysate. Avoid plates that have completely cleared lawns, where all bacteria have been eliminated, so that no opaque patches are visible.
15. Be careful not to expose the bottom of the filter to contamination (keep it in the plastic half in which it was sealed).
16. Do not store stocks at −80°C, as storage at −20°C is sufficient to ensure their survival for more than 10 years.

Acknowledgments

Authors would like to thank the "Conselho Nacional de Desenvolvimento Científico e Tecnológico" (CNPq; Grants #303384/2022-0, #311863/2022-1, #402745/2021-3) and "Coordenação de Aperfeiçoamento de Pessoal de Nível Superior" (CAPES-Brazil; Finance Code 001; UFPB PrInt 88881.311776/2018–01).

References

1. Burgos G, Ambuludí A, USFQ SARS-CoV-2 Consortium, Morales-Jadán D, Garcia-Bereguiain MA, Muslin C, Armijos-Jaramillo V (2023) A tool for the cheap and rapid screening of SARS-CoV-2 variants of concern (VoCs) by Sanger sequencing. Microbiol Spectr 11:e05064-22
2. Zupanc M, Zevnik J, Filipić A, Gutierrez-Aguirre I, Ješelnik M, Košir T, Ortar J, Dular M, Petkovšek M (2023) Inactivation of the enveloped virus Phi6 with hydrodynamic cavitation. Ultrason Sonochem 95:106400
3. European Food Safety Authority (2023) Covid-19, https://www.efsa.europa.eu/en/topics/covid-19 (accessed 10 Feb 2025)
4. Gharibzahedi SMT, Altintas Z (2024) State-of-the-art sensor technologies for tracking SARS-CoV-2 in contaminated food and packaging: towards the future techniques of food safety assurance. Trends Anal Chem 170: 117473
5. Wathore R, Gupta A, Bherwani H, Labhasetwar N (2020) Understanding air and water borne transmission and survival of coronavirus: insights and way forward for SARS-CoV-2. Sci Total Environ 749:141486
6. Chen Y, Zhao S, Xu Y, Cai M, Zhang G (2023) SARS-CoV-2 transmission via maritime cold chains: a statistical analysis of nucleic acid detection results of cold chain food imported from Fuzhou ports. Heliyon 9:11
7. da Silva RT, dos Santos Franco AJ, de Souza Grilo MM, Lima A, Saraiva KLA, Carvalho RSF, Souza Pedrosa GT, Shaffner DW, Magnani M (2025) SARS-CoV-2 surrogate bacteriophage φ6 cross-contamination between fruits and gloves, survival on discarded gloves and inactivation by photodynamic treatment. Food Microbiol 125:104645
8. Jung S, Kim DH, Ahn HS, Go HJ, Wang Z, Yeo D, Woo S, Seo Y, Hossain MI, Choi IS, Ha SD, Choi C (2023) Stability and inactivation of SARS-CoV-2 on food contact surfaces. Food Control 143:109306
9. da Silva RT, de Souza Grilo MM, Pimentel TC, Lucena FA, Schaffner DW, Souza Pedrosa GT, Magnani M (2024) An overview of food-borne viruses and SARS-CoV-2 in foods and

food-contact surfaces: survival, transfer, surrogates use, and mathematical modeling. Curr Opin Food Sci 55:101119
10. Filipić A, Primc G, Zaplotnik R, Mehle N, Gutierrez-Aguirre I, Ravnikar M, Mozetič M, Žel J, Dobnik D (2019) Cold atmospheric plasma as a novel method for inactivation of potato virus Y in water samples. Food Environ Virol 11:220–228
11. McMinn BR, Ashbolt NJ, Korajkic A (2017) Bacteriophages as indicators of faecal pollution and enteric virus removal. Lett Appl Microbiol 65:11–26
12. Chhibber S, Kaur P, Gondil VS (2018) Simple drop cast method for enumeration of bacteriophages. J Virol Methods 262:1–5
13. Abedon ST, Yin J (2009) Bacteriophage plaques: theory and analysis. In: Clokie Martha RJ, Kropinski Andrew M (eds) Springer protocols, 1st edn. Humana Totowa, New Jersey
14. Ács N, Gambino M, Brøndsted L (2020) Bacteriophage enumeration and detection methods. Front Microbiol 11:594868
15. Cormier J, Janes M (2014) A double layer plaque assay using spread plate technique for enumeration of bacteriophage MS2. J Virol Methods 196:86–92
16. Kropinski AM, Mazzocco A, Waddell TE, Lingohr E, Johnson RP (2009) Enumeration of bacteriophages by double agar overlay plaque assay. In: Clokie Martha RJ, Kropinski Andrew M (eds) Springer protocols, 1st edn. Humana Totowa, New Jersey
17. Blanch AR, Lucena F, Muniesa M, Jofre J (2020) Fast and easy methods for the detection of coliphages. J Microbiol Methods 173:105940
18. Peng X, Nguyen A, Ghosh D (2018) Quantification of M13 and T7 bacteriophages by TaqMan and SYBR green qPCR. J Virol Methods 252:100–107

Check for updates

Chapter 15

Detection of Human Norovirus and Hepatitis A Virus in Berries

Éric Jubinville, Marie-Ève Collard, Anne-Marie Lauzier and Julie Jean

Abstract

The protocol described in this chapter focuses on the detection of foodborne viruses, specifically human norovirus and hepatitis A virus in berries. Viruses are eluted from the berries and precipitated using PEG/NaCl. Chloroform-butanol is also used for purification. RNA extraction is based on disruption of the viral capsid using a chaotropic lysis buffer and silica beads to recover the RNA. RT-qPCR is used to detect the viral genome.

Keywords Human norovirus, Hepatitis A virus, Berries, PEG precipitation, Molecular detection, RT-qPCR

1 Introduction

Human norovirus (HuNoV) and hepatitis A virus (HAV) are key causes of foodborne viral illnesses in humans [1]. For detecting HuNoV infectious particles, cell culture techniques are extremely tedious and costly [2]. Detection of HAV in food matrices also requires a method better suited to routine use. In both cases, the reverse-transcriptase real-time polymerase chain reaction (RT-qPCR) is preferred. Since berries contain substances that can inhibit RT-qPCR, the virus genome extraction method must provide very clean RNA. This is achieved by combining viral particle elution with agitation followed by precipitation with polyethylene glycol and salt then extracting RNA from capsids disrupted using chaotropic reagents and adsorbing it to silica particles. RT-qPCR coupled with a fluorescent specific probe allows measurement of the amplified target template concentration. Controls are included to allow correct calculation of the viral concentration in the sample analyzed. The protocol described in this chapter is based on ISO standard 15216:1 for detection of viral RNA in berries [3].

Marciane Magnani and Geany Targino de Souza Pedrosa (eds.), *Detection and Quantification of Bacteria, Yeast, Viruses, and Protozoan in Foods and Freshwater*, Methods and Protocols in Food Science, https://doi.org/10.1007/978-1-0716-5064-6_15,

2 Materials

2.1 General Reagents Used as Supplied

Molecular-biology-grade water, polyethylene glycol (PEG-8000), sodium chloride (NaCl), potassium chloride (KCl), disodium hydrogen phosphate (Na_2HPO_4), potassium dihydrogen phosphate (KH_2PO_4), Tris base, glycine, beef extract powder, pectinase from *Aspergillus aculeatus*, chloroform, *n*-butanol, sodium hydroxide (NaOH; 10 M), hydrochloric acid (HCl; 5 M), ethylenediaminetetraacetic acid (EDTA) disodium dihydrate.

2.2 Reagents Needing Preparation

2.2.1 5× PEG/NaCl Solution (see Note 1)

Sterilize by autoclaving, store at room temperature for up to 6 months.

2.2.2 Chloroform-Butanol

Combine 100 mL ± 10 mL of chloroform and 100 mL ± 10 mL of n-butanol. Store in an amber glass bottle at room temperature for up to 12 months.

2.2.3 Phosphate-Buffered Saline (PBS)

Dissolve 8 g ± 0.1 g of sodium chloride, 0.2 g ± 0.01 g of potassium chloride, 1.15 g ± 0.01 g of disodium hydrogen phosphate and 0.2 g ± 0.01 g of potassium dihydrogen phosphate in 1000 mL ± 10 mL of water. Adjust the pH to 7.3 ± 0.2, sterilize by autoclaving then dispense 50 mL aliquots into Falcon tubes (50 mL) in a biological safety cabinet. Store at room temperature for up to 6 months.

2.2.4 Tris/Glycine/Beef Extract (TGBE) Buffer

Dissolve 12.1 g ± 0.2 g of Tris base, 3.8 g ± 0.1 g of glycine, 10 g ± 1 g of beef extract powder in 1000 mL ± 10 mL of water. Adjust pH to 9.5 ± 0.2, sterilize by autoclaving, dispense 40 mL aliquots into Falcon tubes (50 mL) in a biological safety cabinet. Store at 4°C for up to 6 months.

2.2.5 Tris Solution

Dissolve 12.1 g ± 0.2 g of Tris base in 90 mL ± 1 mL of water. Adjust pH to 8.0 ± 0.2 and bring the volume to 100 mL ± 1 mL with water. Sterilize by autoclaving and store at 4°C for up to 6 months.

2.2.6 EDTA Solution

Dissolve 18.6 g ± 0.2 g of EDTA disodium dihydrate in 90 mL ± 1 mL of water. Adjust the pH to 8.0 ± 0.2 and bring the volume to 100 mL ± 1 mL with water. Sterilize by autoclaving and store at 4°C for up to 6 months.

2.2.7 Tris-EDTA Solution

Combine 1000 µL ± 10 µL of Tris solution and 200 µL ± 10 µL of EDTA solution with 100 mL ± 1 mL of water. Store at 4°C for up to 6 months.

2.2.8 Viral Processing Control

Under air with 5% CO_2, grow Mengo virus strain MC_0 (Spanish Type Culture Collection or equivalent, *see* **Note 2**) on a confluent monolayer of the HeLa cell line (ATCC CCL-2 or equivalent) until cytopathic effects can be observed (~72 h). Recover the virus using freezing-thawing cycles. Centrifuge the culture liquid at 3000 × *g* for 10 min to pellet the cell debris and recover the supernatant containing the virus. Store at −80°C.

2.3 Extraction Reagents

Silica, lysis solution (chaotropic buffer containing guanidine thiocyanate), wash and elution buffers for extraction of viral RNA (e.g., Nuclisens – bioMérieux, *see* **Note 3**).

2.4 RT-qPCR Reagents

2.4.1 HAV primers and probe: HAV68 fw TCA CCG CCG TTT GCC TAG; HAV240 rev GGA GAG CCC TGG AAG AAA G; HAV150(-) probe: CCT GAA CCT GCA GGA ATT AA (5′-FAM and 3′ MGBNFQ).

2.4.2 Norovirus GI primers and probe: QNIF4 fw CGC TGG ATG CGN TTC CAT; NV1LCR rev CCT TAG ACG CCA TCA TCA TTT AC; TM9 probe: TGG ACA GGA GAT CGC (5′-FAM and 3′-TAMRA).

2.4.3 Norovirus GII primers and probe: QNIF2 fw ATG TTC AGR TGG ATG AGR TTC TCW GA; COG2R rev TCG ACG CCA TCT TCA TTC ACA; QNIFs probe AGC ACG TGG GAG GGC GAT CG (5′-FAM and 3′-TAMRA).

2.4.4 Mengo virus primers and probe: Mengo 110 fw GCG GGT CCT GCC GAA AGT; Mengo 209 rev GAA GTA ACA TAT AGA CAG ACG CAC AC; Mengo 147 probe ATC ACA TTA CTG GCC GAA GC (5′-FAM and 3′-MGBNFQ)

2.4.5 RT-qPCR one-step amplification kit that can be run in 20 μL: 15 μL incl. reaction mix, primers and probe, enzyme mix, water and reference dye (if necessary) plus 5 μL of extracted RNA sample. The Bio-Rad one-step iTaq universal probe kit is used in the present protocol.

2.4.6 RT-qPCR thermal cycler equipped with an energy source suitable for the excitation of fluorescent molecules, plus optical detection system for real-time detection of fluorescence signals generated during RT-qPCR with hydrolysis probe chemistry (ABI7500 is used in this protocol).

2.5 Equipment

Micropipettes and aerosol-resistant tips (1000 μL, 200 μL, 20 μL, 10 μL), pipette filler and pipettes (25 mL, 10 mL, 5 mL), vortex mixer, refrigerator (4°C), shaker (50 oscillations min^{-1}), incubator/shaker (37°C, approx. 320 oscillations min^{-1}), rocking platform (with room T and 4°C settings, approx. 60 oscillations min^{-1}), aspirator or equivalent apparatus for removing supernatant, water bath (60°C), centrifuge and rotor (10,000 × *g*, at 5°C, 35 mL capacity; 10,000 × *g* at 5°C, chloroform-resistant tubes with 2 mL capacity; 4000 × *g* at room T for centrifugal filter concentration devices), centrifuge tubes and bottles (1.5 mL, 5 mL, 15 mL, 50 mL), chloroform-resistant tubes with 2 mL capacity, pH meter or pH testing strips, mesh filter bags (400 mL), 48-well deep-well micro-assay plates, RNA extraction material (silica and chaotropic agent), Integra pipette, RT-qPCR thermal cycler (e.g., ABI7500 real-time detection of fluorescence signals for hydrolysis probe chemistry), associated consumables for RT-qPCR, such as the Bio-Rad one-step iTaq universal probe kit.

3 Methods

3.1 Elution of the Virus from the Food Sample

3.1.1 The number of samples that can be processed is limited to the number of places in the centrifuge head. This protocol is based on 6 samples at a time including one negative processing control. However, it is possible to start a second ISO protocol with a delay of 30 min between sample runs. A trained technician normally can process 2 sets of 6 samples in about 5 h. A general overview of the method is illustrated below (Fig. 1).

3.1.2 Weigh 25 g ±0.3 g of soft fruit (do not chop; *see* **Note 4**) and transfer the sample to a 400-mL mesh filter bag under a laminar flow hood.

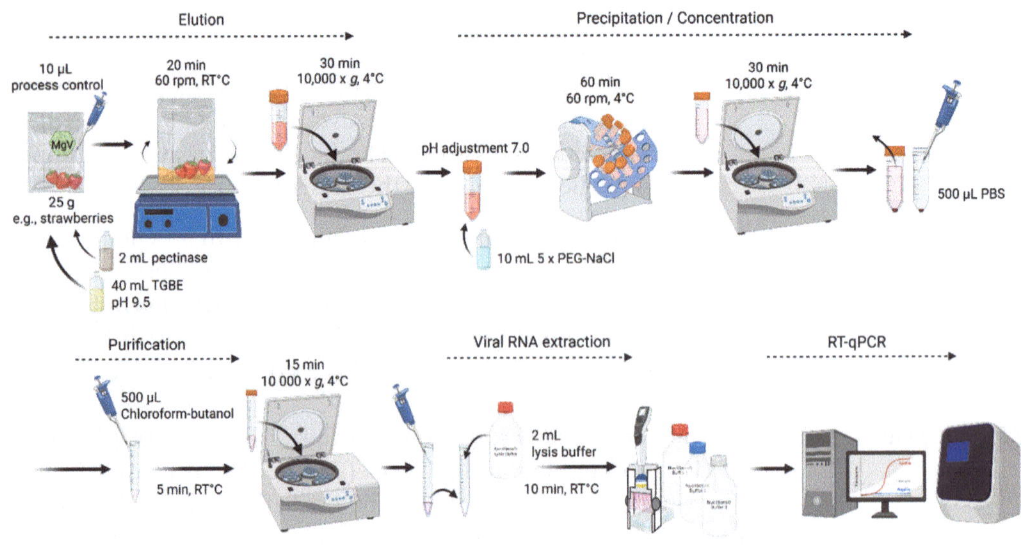

Fig. 1 Method of virus detection in berries, based on ISO standard 15216:1. Viral particle elution, precipitation, concentration, purification, RNA extraction, and RT-qPCR are described in the text

3.1.3 Add 10 μL ± 0.5 μL of process control virus (Mengo virus) to each sample.

3.1.4 Add 40 mL ± 1 mL of TGBE and 1140 units of pectinase from *Aspergillus aculeatus* to the buffer (*see* **Note 5**).

3.1.5 Maintain constant rocking motion at approximately 60 oscillations min^{-1} for 20 min ± 1 min at room temperature.

3.1.6 Monitor the pH at 10-min intervals during pectin lysis. If the pH falls below 9.0, it must be adjusted to 9.5 ± 0.5 with NaOH (10 M). Extend the treatment time by 10 min each time the pH is adjusted. Do not adjust the pH more than three times per sample.

3.1.7 Decant the eluate from the filtered compartment into a centrifuge tube.

3.1.8 Clarify by centrifugation at 10,000 × *g* for 30 min ± 5 min at 4°C.

3.2 Precipitation and Concentration of the Virus

3.2.1 Decant the supernatant into a single clean tube and adjust to pH 7.0 ± 0.5 with HCl (5 M, *see* **Note 6**).

3.2.2 Add 0.25 volume of 5× PEG/NaCl solution (*see* **Note 7**) to obtain final concentrations of 100 g/L PEG and 0.3mol/L NaCl, shake for 60 s ± 5 s to homogenize then hold at 4°C for 60 min ± 5 min with rocking motion at approximately 60 oscillations min^{-1}.

3.2.3 Centrifuge at 10,000 × *g* for 30 min ± 5 min at 4°C.

3.2.4 Decant and discard the free supernatant, centrifuge the loose pellet at 10,000 × *g* for 5 min ± 1 min at 4°C to compact it.

3.2.5 Discard the supernatant; place the Falcon tube inverted on absorbent paper to dewater the pellet; resuspend the pellet in 1000 μL of PBS.

3.3 Purification of the Virus

3.3.1 Transfer the suspension to a chloroform-resistant centrifuge tube.

3.3.2 Add 1000 μL ± 20 μL of chloroform/butanol, vortex mix for 30 s then hold at room temperature for 5 min (*see* **Note 8**).

3.3.3 Centrifuge at 10,000 × *g* for 15 min ± 1 min at 4°C.

3.3.4 Carefully transfer the aqueous (upper) phase to a fresh tube and retain for RNA extraction (*see* **Note 9**).

3.3.5 Process samples immediately or store at 4°C for no more than 24 h.

3.4 Extraction of Viral RNA

3.4.1 Extract RNA from the entire sample using an appropriate guanidine thiocyanate lysis buffer and silica adsorption-based method.

3.4.2 Elute purified RNA into 100 μL of elution buffer and retain for RT-qPCR analysis.

3.4.3 Extracted RNA should be processed immediately or stored at 4°C for no more than 24 h or at −80°C for longer periods.

3.4.4 For each batch of samples tested, a negative RNA extraction control (water) must be included.

3.4.5 Sample preparation using the Nuclisens extraction method

3.4.5.1 In a biosafety level 2 cabinet, add 2000 μL of lysis buffer to each sample (3.1.18), mix by vortex and incubate at room temperature for 10 min.

3.4.5.2 Centrifuge at 1800 × *g* for 2 min at 4°C.

3.4.5.3 Pre-warm the thermo-shaker to 80°C.

3.4.6 Plate preparation

3.4.6.1 Prepare the deep-well micro-assay plate manually and identify each well (samples in line 1) and each row (buffer 1–3).

3.4.6.2 Dispense 2000 μL of buffer 1 into each well of lines 2 and 3.

3.4.6.3 Dispense 2000 μL of buffer 2 into each well of lines 4 and 5.

3.4.6.4 Dispense 2000 μL of buffer 3 into each well of line 6.

3.4.6.5 Prepare the elution tubes by adding 100 μL of buffer 3 to the PCR strip tubes.

3.4.6.6 Transfer all samples to line 1 of the micro-assay plate. Include the negative extraction control (water).

3.4.6.7 Vortex the magnetic silica beads and add 50 μL to each well in line 1.

3.4.7 Pipette preparation

3.4.7.1 Attach the appropriate Grip Tips (1000 μL) to the Integra pipette and ensure they are properly sealed.

3.4.7.2 Place the Integra pipette on the eGENE-UP holder.

3.4.7.3 Block the eGENE-UP holder in the upper position.

3.4.7.4 Slide the deep-well micro-assay plate under the eGENE-UP holder and lower the holder slowly to wet the Grip Tip points in line 1.

3.4.8 Starting the pipette protocol

3.4.8.1 From the main menu, using the touch wheel of the Integra pipette, select Custom and choose Extraction V01.

3.4.8.2 Press RUN to start the program.

3.4.9 RNA binding and washing on silica beads

3.4.9.1 Follow the indication on the Integra pipette; after mixing the magnetic particles into the sample, the program will display a prompt to insert the magnet, then press RUN again (*see* **Note 10**).

3.4.9.2 At the end of the cycle (beeping sound), pull the holder and the magnetic bar up to the raised position using the wheel then shift the pipette to line 2 (lower the tips into the liquid using the wheel).

3.4.9.3 Press RUN.

3.4.9.4 At the end of the cycle (beeping sound), pull the holder and the magnetic bar up to the raised position using the wheel then shift the pipette to line 3 (lower the tips into the liquid using the wheel).

3.4.9.5 Press RUN.

3.4.9.6 At the end of the cycle (beeping sound), pull the holder and the magnetic bar up to the raised position using the wheel then shift the pipette to line 4 (lower the tips as described above).

3.4.9.7 Press RUN, wait for the beep, and release the magnet.

3.4.9.8 Press RUN, wait for the beep, and insert the magnet.

3.4.9.9 Press RUN.

3.4.9.10 At the end of the cycle (beeping), pull the holder and the magnetic bar up to the raised position using the wheel then shift the pipette to line 5 (lower the tips as described above).

3.4.9.11 Press RUN.

3.4.9.12 At the end of the cycle (beeping), pull the holder and the magnetic bar up to the raised position using the wheel then shift the pipette to line 6 (lower the tips as described above).

3.4.10 Recovery of RNA in the elution tube

3.4.10.1 Press RUN.

3.4.10.2 At the end of the cycle (beeping sound), with one hand, hold the magnetic bar in the lower position; with the other hand, pull the holder slowly upward such that the magnetic beads stay near the magnet. Raise the holder to the upper position and carefully check that the magnetic beads are now at the bottom of the Grip Tips.

3.4.10.3 Press RUN (the program will flush the liquid).

3.4.10.4 Remove the deep-well micro-assay plate and place the elution tube (qPCR strip) previously filled with buffer 3.

3.4.10.5 Press RUN and release the magnet.

3.4.10.6 Press RUN and lower the tips into the qPCR strip tubes using the wheel.

3.4.10.7 Press RUN; when the beep sounds, raise the Grip Tips slightly above the elution tube using the wheel.

3.4.10.8 Press RUN and wait for the last beeping sound.

3.4.10.9 Raise the holder to the upper position, cap each qPCR strip tube.

3.4.11 RNA elution from silica beads

3.4.11.1 Place the qPCR strip in the thermo-shaker and shake at 1400 oscillation min^{-1} at 80°C for 10 min.

3.4.11.2 Centrifuge the qPCR strip (quick spin).

3.4.11.3 Place the qPCR strip on the elution magnetic holder.

3.4.11.4 Allow the silica beads to collect on the magnet (~30 s) then transfer the RNA to a clean qPCR strip.

3.4.11.5 Place the strip on ice for immediate use in RT-qPCR or freeze at −80°C for storage.

3.5 RT-qPCR Viral Detection Generalities

3.5.1. Primers and probes must be purchased (e.g., IDT DNA).

3.5.2 Several RT-qPCR detection kits can be used. However, the reaction total volume must be 20 μL (15 μL of mix and 5 μL of RNA).

3.5.3 Several controls must be included in the 96-well plate, such as negative RT-qPCR control (water) and negative RNA extraction control (water).

3.5.4. Inhibition is measured using the cycle quantification method (ΔCq). The difference between the undiluted and diluted RNA samples must be at least 3.33. If ΔCq is lower than 3.33, a PCR inhibitor was present, and the diluted sample Cq must be used for analysis.

3.5.5 All samples must be analyzed at least in duplicate.

3.5.6 To reduce cost and reagent waste, run the Mengo virus RT-qPCR first; if all samples are valid, run the targeted foodborne virus. The ISO method requires at least 1% Mengo virus recovery in each sample tested.

3.6 RT-qPCR Plate Preparation Using Bio-Rad One-Step iTaq Universal Probe Kit

3.6.1 Thaw the iTaq universal probe reaction mix, the primers and probe on ice and place the iScript reverse transcriptase and samples on ice.

3.6.2 Mix thoroughly all tubes and quick centrifuge.

3.6.3 The volumes per sample are as follows: 10 μL of iTaq universal probe reaction mix, 2.625 μL of RNAse free water, 0.625 μL (625 nM) of forward primer, 0.625 μL (625 nM) of reverse primer, 0.625 (625 nM) μL of probe, 0.5 μL of reverse transcriptase. The total master mix volume is 15 μL.

3.6.4 Prepare enough (e.g., 4 extra wells) master mix to ensure adequate volume for all samples and controls; mix thoroughly and quick centrifuge.

3.6.5 Dispense 15 μL of master mix to each well to be used for RT-qPCR.

3.6.6 Dilute viral target RNA 1/10 using Nuclisens buffer 3.

3.6.7 Dilute extracted Mengo virus RNA 1/10, 1/100, and 1/1000 using Nuclisens buffer 3 or RNAse-free water (process controls).

3.6.8 Load wells with 5 μL of extracted process control RNA (*see* **Note 11**).

3.6.9 Load wells with 5 μL of RNA sample, undiluted and diluted 1/10 (3.4.11.5).

3.6.10 Load wells with 5 μL of negative extraction control (water).

3.6.11 Load wells with 5 μL of negative RT-qPCR control (water).

3.6.12 Seal the plate with an adhesive film.

3.6.13 Vortex-mix then quick-centrifuge the 96-well micro-assay plate.

3.6.14 Place the plate in your RT-qPCR thermal cycler.

3.7 Reading the RT-qPCR Results Using the ABI7500

3.7.1 Open the application software; under Assay, select Standard Curve (Absolute Quantification)

3.7.2 Prepare the plate layout accordingly by selecting the target (make sure you select the associated probe).

3.7.3 Use the following thermal cycling protocol: Stage 1 (reverse transcription) 50°C for 10 min; Stage 2 (polymerase activation and cDNA denaturation) 95°C for 3 min; Stage 3 Step 1 (denaturation) 95°C for 15 s; Stage 3 Step 2 (annealing, extension and plate fluorescence read) 60°C for 60 s. Repeat stage 3 steps 1 and 2 for 45 cycles.

3.7.4 Press Start Run.

3.7.5 Analyze the data. Refer to Fig. 2 for an example of the amplification plot generated during the detection of HuNoV.

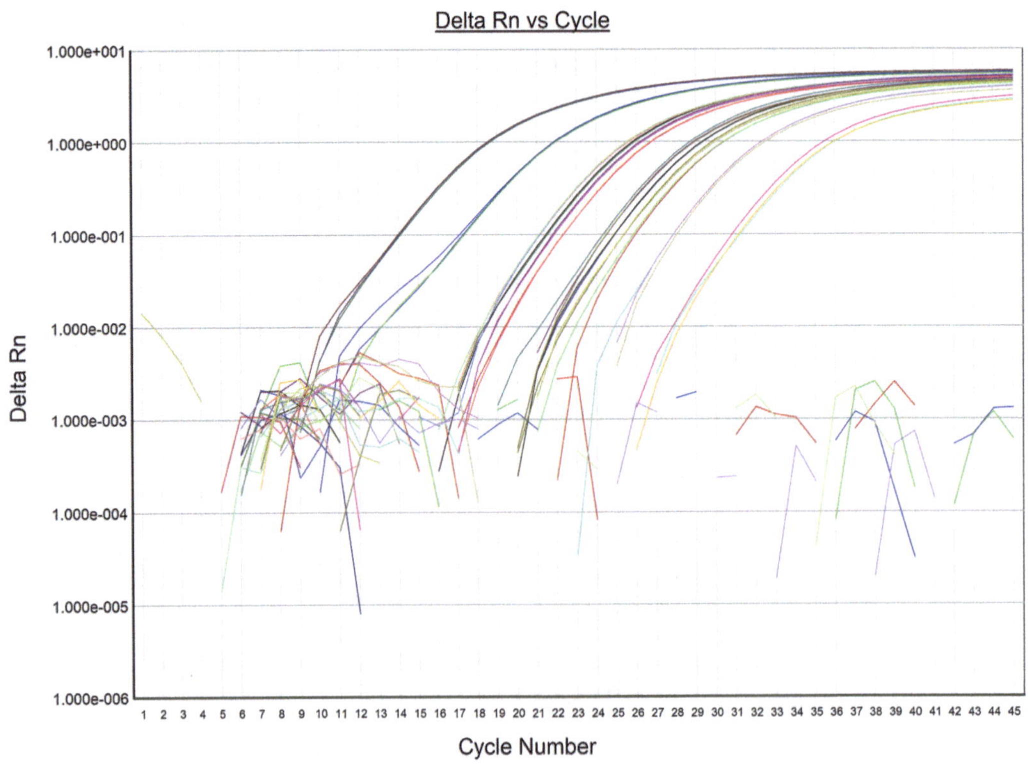

Fig. 2 Quantification plot of human norovirus detection using the ABI7500. Red arrows represent the standard curve, Green arrow represent a positive sample and the Blue arrow represent the negative control sample

4 Notes

1. Dissolve the salt completely in 200 mL of water with magnetic stirring. Add the remaining water and PEG. Stir manually until the magnetic stirring bar moves freely, then alternate between manual stirring and magnetic stirring to ensure thorough dissolution. Heat for a few minutes at low temperature. Do not overheat. Wait until the solution becomes transparent (may take several hours) before sterilizing (sterilize on the same day). The sterilized solution may separate. To homogenize, invert the container several times every 15 min for 1–1.5 h. PEG solution may precipitate over time. A water bath set at 37°C may be used to overcome this problem.
2. Although Mengo virus is named in the ISO standard, murine norovirus 1 may be used, which is often easier to obtain and culture.
3. Several RNA extraction kits can be used. However, the ISO standard requires that the kit be silica-based with a guanidine thiocyanate chaotropic lysis buffer.
4. When testing fresh strawberries, the calyx must be removed.
5. Pectinase from *Aspergillus niger* is acceptable. Use at least 30 units.
6. Count the number of drops added to the first sample to adjust the pH to 7, to speed up the process for subsequent samples of the same food matrix.
7. Add the PEG to the Falcon tube gently, with the pipette tip touching the inner wall. Re-draw liquid into the pipette and expel 1–2 times to ensure total addition.
8. When processing two batches of samples, wait until the centrifugation (Step 3.1.8) time remaining is 7 min before adding chloroform-butanol to the 2nd batch, to make sure that holding time with the solvent is limited to 5 min.
9. Recover the sample (approximately 1 mL) without disturbing the organic phase. Proceed very gently.
10. Visually check that the silica beads are captured by the magnet.
11. To validate the standard curve (Mengo virus, 3-point, run in duplicate), the ISO standard recommends at least 1% recovery of Mengo virus in all samples to consider any targeted viral RT-qPCR data. If Mengo virus recovery is lower, the sample must be reprocessed, and which means repeating all steps from 3.1.2.

References

1. World Health Organization (2023) Joint FAO/WHO Expert Meeting on microbiological risk assessment of viruses in foods Part 1: food attribution, analytical methods, and indicators, https://www.who.int/publications/m/item/jemra-of-viruses-in-foods-part1-food-attribution-analytical-methods-and-indicators
2. Bosch A et al (2018) Foodborne viruses: detection, risk assessment, and control options in food processing. Int J Food Microbiol 285:110–128
3. Anonymous (2017) Microbiology of the food chain – horizontal method for determination of hepatitis A virus and norovirus using real-time RT-PCR. Part 1: method for quantification – ISO 15216-1:2017. https://www.iso.org/standard/65681.html

Subject Index

Marciane Magnani and Geany Targino de Souza Pedrosa (eds.), *Detection and Quantification of Bacteria, Yeast, Viruses, and Protozoan in Foods and Freshwater*, Methods and Protocols in Food Science, https://doi.org/10.1007/978-1-0716-5064-6,

D

E

F

G

H

I

K

L

M

MIX
Papier aus verantwortungsvollen Quellen
Paper from responsible sources
FSC® C105338
FSC
www.fsc.org

If you have any concerns about our products,
you can contact us on
ProductSafety@springernature.com

In case Publisher is established outside the EU,
the EU authorized representative is:
Springer Nature Customer Service Center GmbH
Europaplatz 3, 69115 Heidelberg, Germany

Printed by Libri Plureos GmbH
in Hamburg, Germany